Introduction to engineering design

with graphics and
design projects

Prentice-Hall, Inc., Englewood Cliffs, New Jersey 07632

TERRY E. SHOUP
Texas A & M University

LEROY S. FLETCHER
Texas A & M University

EDWARD V. MOCHEL
University of Virginia

Introduction to engineering design

with graphics and design projects

Library of Congress Cataloging in Publication Data

Shoup, Terry E.
  Introduction to engineering design with graphics and design projects.

  Bibliography: p.
  Includes index.
  1. Engineering design. I. Fletcher, Leroy S.,
joint author. II. Mochel, Edward V., joint
author. III. Title.
TA174.S56   620'.00425   80–23798
ISBN 0–13–482364–8

Editorial production supervision
and interior design by: James M. Chege

Cover design by: Edsal Enterprises

Manufacturing buyer: Anthony Caruso

Printed in the United States of America

10 9 8 7 6 5 4 3 2 1

PRENTICE-HALL INTERNATIONAL, Inc., *London*
PRENTICE-HALL of Australia Pty. Limited, *Sydney*
PRENTICE-HALL of Canada, Ltd., *Toronto*
PRENTICE-HALL of India Private Limited, *New Delhi*
PRENTICE-HALL of Japan, Inc., *Tokyo*
PRENTICE-HALL of Southeast Asia Pte. Ltd., *Singapore*
WHITEHALL BOOKS LIMITED, *Wellington, New Zealand*

To
Jennifer and Matthew
Laura and Daniel
Martha and Fritz

Contents

4 Design considerations 77

5 Engineering communications 101

6 Dimensions and unit systems 129

7 Engineering graphics 155

With the increasing importance of technological developments in everyday life, the importance of engineers and scientists has also increased. As problems become more complex and diverse, the engineer and scientist must develop skills in problem solving and design which are independent of the type of problem considered. In order to equip themselves to handle effectively the problems they confront, engineers must develop an understanding of the engineering design process as well as techniques for applying the design process to real world problems.

This textbook has been prepared as an introductory textbook for first-year students in engineering and engineering technology, to provide some background and practice in engineering design. The book introduces the fundamental concepts of the design process and offers instruction in the formulation of design problems, techniques for innovation and creativity in engineering design, model building, and appropriate design considerations, communications, engineering graphics, the use of standard commercial parts, and systems design and project planning, as well as specific engineering design projects for students.

Because of its basic nature, the text does not treat its subjects in great depth. References are provided, however, for students who would like to pursue any of the subject areas in more detail. Problems are provided at the end of appropriate chapters to give the student practice in applying the skills that have been learned.

Preface

A firm understanding of the engineering design process is developed through the use of practice engineering design.

The material in the text is divided into three sections, as it might be presented in an undergraduate course. In the first section, Chapter one describes the overall engineering design process and introduces the steps necessary for implementation of this process in real world problems. Chapter two describes the thought processes associated with the development of innovative ideas in engineering. Chapter three describes both experimental and analytical techniques for building models appropriate to the design process. Chapter four discusses the various aspects of engineering design, including standards, laws, regulations, economics, ethics, and professional responsibilities.

The second section of the book deals with engineering communication including graphical and written materials. Chapter five discusses the development of proposals, engineering reports, oral reports, and other communication techniques. Chapter six introduces the International System of Units (SI) and discusses the base units and derived units as well as their use in engineering problems. Dimensional analysis is also reviewed. Chapter seven introduces the basic concepts of graphics, including sketching, lettering, projection systems, and standard conventions and practice.

The third section of the textbook provides a discussion of design projects and their components. Chapter eight reviews the types of standard parts available for engineering design, and techniques for their selection and use. Chapter nine discusses project planning and the technique of systems design for implementation of a design project. Chapter ten provides a brief analysis of a number of engineering design projects appropriate to the beginning engineering student. The Appendices for this textbook have been selected to provide the resource materials for the development of design projects. Specific appendices include fasteners, materials, engineering symbols, metric units (SI), anthropometric data, details on areas and circumferences, human factors, and material cost information.

Throughout the development of this book, a number of people contributed constructive ideas, and provided useful material. Thanks are due to the American Society for Mechanical Engineers for providing various ANSI standards and for permission to use them in this text. Particular thanks for assistance in reviewing sections of this manuscript are due many people, especially L. Sanchez, J.E. Cox, I. Soudek, C.E.G. Przirembel, A.D. Gianniny, L. Harrisberger, M.R. Green, K. Wessely, D.R. Reyes-Guerra, and L. Rader.

TERRY E. SHOUP
LEROY S. FLETCHER
EDWARD V. MOCHEL

Introduction to engineering design

with graphics and
design projects

Oil refinery. (*Courtesy* Exxon USA.)

Engineering design

1

Engineering is the profession in which a knowledge of the mathematical and natural sciences gained by study, experience, and practice is applied with judgment to develop ways to utilize, economically, the materials and forces of nature for the benefit of mankind.* Engineering technology is that part of the technological field which requires the application of scientific and engineering knowledge and methods combined with technical skills in support of engineering activities; it lies in the occupational spectrum between the craftsman and the engineer at the end of the spectrum closest to the engineer.* Simply stated, engineering design is the physical and mental activity associated with the creation of products and processes to meet changing needs. It is the primary activity of the engineering professions and is practiced by both engineers and engineering technologists. Engineering design combines the mental activity of idea generation and evaluation with the physical activity of communication. It is an exciting and challenging function for men and women who wish to apply their talents to find solutions to the problems that face our world.

Ever since the beginning of civilization, people have been systematic and innovative. Long before the term *engineer* was coined, the basic elements of engineering design were used to create ways to improve life and to harness the environment. No one can say for sure when the process of engineering design actually began, however the basic thought process of design is older than language itself. Carvings on the walls of caves serve as evidence that their early human inhabitants not only generated ideas but also utilized creative talents to communicate these thoughts and concepts.

The motivation for creative activity is as basic as survival itself. Recognizing the need to secure adequate nourishment with the minimum of effort, people first invented ways to improve hunting and farming. These initial responses to the basic need for food and clothing began the technological evolution we see continuing in the food and fiber production industries of today. The need for transportation of heavy objects resulted in the invention of the wheel. This fundamental discovery initiated the technological evolution we see continuing in the transportation systems we use and enjoy today. People have always had a need for shelter in order to live safely and comfortably in a changing environment. The first response to this need was the use of caves and then the design and fabrication of simple tents and wood shelters. These early developments in dwelling design initiated the technological evolution that has led to the modern dwellings that we comfortably occupy today. The logic pattern associated with the process of design is almost a human instinct. Yet it is safe to say that without this natural process, the progress of civilization could not have taken place.

*Engineers' Council for Professional Development, 47th Annual Report, EC9, September 30, 1979.

People have always had a strong drive toward new experiences and the acquisition of new knowledge. As a result, they have become experimenters and developers of countless ideas based on the creative application of materials and the laws of nature. The presence of engineering design and its impact on our lives is evident in almost every facet of our world and its surroundings (Figure 1–1).

The fruits of engineering design allow us to travel faster and farther than any other creature on the earth. They allow us to look at objects thousands of times smaller than any unaided eye can see. The fruits of engineering design allow us to move objects hundreds of times our own weight. They allow us to build objects thousands of times our own size. If properly used, the fruits of design have the potential for enriching our lives in countless ways. If improperly used, the fruits of design have the potential to destroy us. It is an interesting phenomenon that the more one designs, the more one sees that needs to be designed. The result of this paradox is a never-ending spiral of progress.

It is the purpose of this book to explore the process of creation and communication of ideas, with a view toward a better understanding of this important process. For the student of engineering who is just beginning the study of design, an exciting domain awaits. Because of technological advances, students of engineering can now tap technological resources generated over the past centuries. Engineering concepts and designs beyond the wildest dreams of our ancestors are now close to becoming reality in the modern engineering environment (Figure 1–2).

In this chapter we will look first at the reasons for having a rational method for engineering design. We will then look at a basic description of the process of design. We will also look at how this basic process fits into the practice of engineering in the modern industrial environment. Finally, we will conclude with a discussion of how education in design fits into the training of engineering professionals.

1.1   The importance of a rational method of design

The first reason for using a rational method of design is that it helps the engineer to manage his or her task when confronted with a vast amount of input information. Technological knowledge is ever increasing in quantity and scope. It is interesting to note that many of the scientific and technological personnel who have contributed to our present knowledge base have done so within the past several decades. This fact explains the explosion of technological knowledge now taking place. The accelerated production of new knowledge provides an interesting information management challenge to the engineering problem solver. The best way

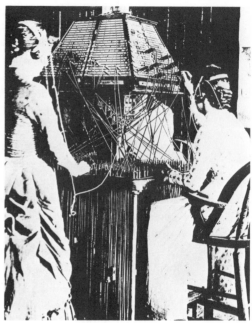

Figure 1–1 Technological progress in the communications field has been possible through the application of human energy and creativity. (*Courtesy* Western Electric).

4

Figure 1–2 The Gossamer albatross was the first human powered flying machine to cross the English Channel. (*Courtesy* Du Pont Company.)

to cope with this challenge is through the use of a systematic method of problem solving. Indeed, experience has shown that if an engineer will use an organized approach to his or her problem solving, many of the difficulties associated with decision making become easier.

A second reason for using a systematic design method is that the use of a common procedure for design greatly facilitates interaction among engineers. Training in the use of a design method provides a common vocabulary for engineers to use in their professional interactions. Most human beings are born with certain mental and physical capabilities for performing design. Yet few untrained persons achieve excellence in engineering. In addition, few engineering achievements today are made by a single engineer working alone. More and more, today's engineers are recognizing the value of team effort in design (Figure 1–3). For this reason a common thought process and a common vocabulary in design are highly desirable. Although the content of each engineering design problem is unique, the methodology for solving these problems is universal and can be described in a specific way.

Figure 1–3 Engineers at work as a team to review the design of a new facility. (*Courtesy* IBM.)

1.2   The design method

Although a number of authorities on the methodology of design have presented descriptions of the process, most of the descriptions tend to be similar. The design process, as we will describe it, involves the six-step procedure diagrammed in Figure 1–4. The steps are represented by boxes on the diagram, and arrows connecting the boxes indicate the sequential order in which they are applied. We will now look at the activity that takes place during each of these steps.

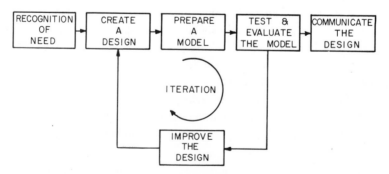

Figure 1–4   The six-step process of design.

Step 1: Recognize the need

Some people mistakenly believe that engineers create need. This is, of course, no more true than the notion that doctors create illness or that farmers create hunger. The products and processes created by engineering design are a direct response to specific needs of society. The logic for this cause and effect relationship is obvious. Engineers cannot make a living in the creation of products and processes for which there is no societal need. The ability to recognize present needs and to anticipate future needs is an extremely valuable talent for an engineer to cultivate (Figure 1–5).

This first step of the design process, the recognition of need, is probably the single most important part of the overall process. Yet it is frequently given inadequate treatment in the thought process of design. The engineering designer needs to make a deliberate effort to focus attention on the preparation of a suitable statement of need for the problem he or she is undertaking. A carefully formulated statement of need can often save considerable time and energy later in the design cycle. Implied in any statement of need is an identification of the real constraints

Figure 1–5 Engineers in the petrochemical industry serve a valuable role in meeting human needs for new chemical products. (*Courtesy* Union Carbide.)

on the problem being considered. A statement of need that recognizes and describes the fundamental elements of a problem can often keep the designer from imposing artifical constraints on his or her design activities. For example, a group of engineers was once engaged in the design of a new type of eyeglasses with expanded visibility and reduced weight. When the design group finally recognized and restated the true need to be that of achieving an improved method of vision correction, they were freed from the artificial constraint of requiring their output to be eyeglasses. The result was the development of the first pair of contact lenses. Until they were able to see the fundamental elements of their design task, they unnecessarily limited their perspective to solutions of low innovation. Some of the most exciting and innovative progress occurs in quantum technological jumps rather than in continuous improvements. For example, automobile manufacturers have worked for years to improve motor oils so that they will need to be replaced less often. A careful assessment of the fundamental elements of this need may reveal that what is actually needed is an engine that does not require a liquid lubricant. A quantum jump in technology may be possible if design efforts are diverted from motor oil design to the design of the engine itself.

It is said that the only way to know that progress has taken place is to be able to measure it. Thus a statement of need should contain adjectives that imply quantitative and qualitative measures of how well a given design satisfies the need. For example, in the case of the illustration of the contact lens design, the measure of quality is based on reduced weight and enhanced visibility.

Once a statement of need has been established, the designer would be well advised to review the statement periodically during the design process. This review should be used to see if progress is being made in meeting the need and also should provide an opportunity for the designer to see if a clearer statement of need is now apparent.

Step 2: Create a design

Once the need has been clearly recognized and stated in a succinct way the next step is to begin creating design ideas that will satisfy this need. Of all the steps in the design process, this step requires the most ingenuity and imagination. Although every person is born with a certain amount of creative talent, much of our natural creativity is suppressed by our social environment. For example, many people are reluctant to suggest a new idea because they fear that someone will laugh. As a result these people tend to suppress their basic creativity and thus become less and less productive in their work. Fortunately there are ways of overcoming these suppressive influences so that productive, innovative thought can take place. We will discuss some of these important methods in detail in

Chapter 2. The overall objective in the design process is to identify a good idea or combination of ideas to meet the stated need. Not surprisingly, the idea creation step may require a number of attempts in order to achieve an optimum result. Sometimes it works well to optimize by creating completely new ideas, and sometimes it works well to modify existing ones. For example, in the automotive industry new design ideas take the form of modifications of existing equipment. In the toy industry, on the other hand, new design ideas take the form of completely new and different products.

Step 3: Prepare a model

Once an idea has been created, it becomes necessary to find a means to evaluate the quality of that design in satisfying the need requirements. One way to do this, of course, would be to build the suggested design idea. This procedure is usually impractical for reasons of cost, time, and effort. In order to conserve cost, time, and effort, engineers frequently make use of a simplified model to evaluate a design idea. A model may be real or abstract and may be anything from a simple mental image of the idea to an elaborate mathematical or physical reproduction of the proposed concept. For example, an engineer designing a new type of tennis racket might make a model in the form of a simple sketch of the proposed idea. The sketch will show all the important features of the new design and will allow the engineer to make changes or improvements by using only a pencil and an eraser. This kind of model is frequently used in engineering design. There are many other types of models available to the engineer, and each has its own advantages and limitations. The ability to work with a variety of models is a talent that every designer should cultivate. Chapter 3 of this book is devoted to the topic of engineering model preparation.

Step 4: Test and evaluate the model

Once the model has been prepared, it is time to evaluate the proposed design idea by evaluating the performance of the engineering model. In this manner the engineer has a chance to observe how well the proposed design idea satisfies the required need (Figure 1–6).

The testing and evaluation of a model may be very simple or quite involved. For example, in the case of the sketch of the new tennis racket, the testing and evaluation may consist entirely of the mental process of deciding whether the design looks strong enough or if it looks functional enough. If the model is a mathematical formula that describes some important characteristics of the proposed idea, the engineer will likely put numbers into the formula in order to see what quantitative performance the model predicts. For example, in the case of the ten-

Figure 1–6 IBM engineers make noise measurements on a model of a new product in order to verify that the proposed design satisfies the required need. (*Courtesy* IBM.)

nis racket design, if the formula is one that predicts the stresses in the racket handle based on the applied forces, the engineer could use the expected forces in the model to determine if the handle will break due to the stresses. Frequently a model will be mathematically complex. When this happens, the designer may choose to use a computer to assist with the numerical testing. In this case the evaluation of the computer output is still the responsibility of the designer. In the design process there is no substitute for the application of engineering judgment.

Step 5: Improve the design

As a result of the tests performed on the model, the engineer should have a quantitative measure of the success or failure of the idea. The engineer will likely know whether the idea should be abandoned or whether it should be retained for further improvement. One of the fortunate results of the testing and evaluation step is that this process often provides considerable insight into where improvements can and should be made. For example, in the case of the tennis racket design, if the model reveals that the handle will break during normal use, the logical remedy would be to increase the strength of the handle by adding more material at the locations of high stress. It is fortunate that even the poorest design idea will likely give useful information about how to choose a better design. Since a number of different design ideas may be tried, modified, and improved before a final design choice is made, the design cycle of Figure 1–4 tends to be quite iterative. The diagram illustrates this by a curved arrow depicting the iterative flow of this feedback process. In actual practice a single design problem may require from one to a hundred cycles to complete the design process. This suggests that both patience and perseverance are useful attributes for engineers. A well-known illustration of this is to be found in the records of Thomas Edison where he describes trying several hundred ideas before perfecting his design of the light bulb.

Step 6: Communicate the design

No matter how well a design may satisfy a particular human need, it cannot be converted into a useful product or process if the details of the design are not communicated to those who will implement its use (Figure 1–7). The communication step thus links the design cycle to the next stage of engineering activity as the idea moves from conceptualization to utilization. Communication of engineering ideas can be by written words, spoken words, or by pictures, graphs, and drawings. Effective communication is so important to the design process that a significant portion of this text is devoted to the development of skills in this area.

Figure 1–7 A graphical layout of a printed circuit board serves as a communications link between engineering design and manufacturing. (*Courtesy* Woodward Governor Co.)

As we think about applying the engineering design method diagrammed in Figure 1–4, several important facts are worth keeping in mind. First, the human mind is capable of handling straightforward design decisions with remarkable speed. Thus the complete process of making a good engineering design decision need not take excessive time for routine problems. Second, the human mind often has difficulty in defining boundaries between the steps in the design process. Thus the designer may combine one or more of the blocks in the diagram into what seems like a single activity. While this procedure is not wrong, it should be avoided if it results in inadequate treatment of each important step.

1.3   Stages of design

The design process, as described in the previous section, is a thought module that can be applied over and over again in engineering practice. In the industrial environment, design is frequently accomplished through a progressive series of four operational stages. These stages carry the design of a product or process from its inception through to its completion. Each of these four stages will involve a complete design cycle such as the one presented in Figure 1–4. These four stages are connected in series so that the first must be completed before the second can start, and so on. The communications output of the first stage will thus provide a statement of need for the input of the second stage. The activity flow for the four-stage process is shown in diagram form in Figure 1–8. A close inspection of this diagram reveals that it is composed of four design modules labeled "feasibility stage," "preliminary stage," "detail stage," and "revision stage." The important characteristics of each stage will now be discussed.

Stage 1: Feasibility stage

The first stage of the design activity is called the *feasibility stage*. This stage determines whether it is both possible and profitable to undertake a given engineering project. For this reason the need statement will often be phrased "to consider the desirability of . . ." or "to consider the economic feasibility of . . ." The ideas generated during this stage of design usually consist of general statements about overall concepts rather than specific descriptions of hardware. The models for this stage tend to be based on economic theories, market surveys, and the opinions of authorities in the field of interest. For this reason, people from disciplines other than engineering are often part of the design team during the feasibility stage. Their input to the design process is extremely valuable. The output of the test and evaluation block for this stage will be a recommendation either to proceed or to abandon the project. Since much of the material in the feasibility

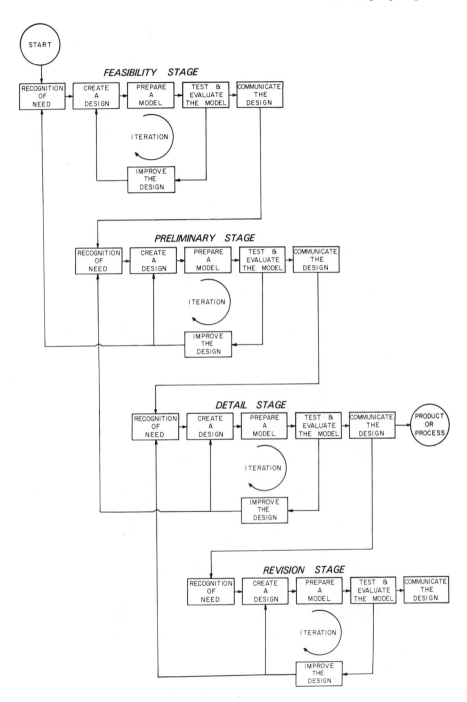

Figure 1–8 The four stages of design.

stage is qualitative rather than quantitative, the output of this stage is usually a written or oral report explaining the facts discovered and presenting a clear justification for the recommendation made. Naturally, if the recommendation is that the project be abandoned, no further design activity takes place until the project is reinitiated. This would likely happen only if new technological conditions or economic facts were found that alter the original conclusion. On the other hand, if the recommendation is to proceed, the next stage of design would begin.

Stage 2: Preliminary stage

The next stage is called the *preliminary stage,* and its purpose is to make qualitative judgments about the types of components and resources to use in meeting the conceptual need identified by the feasibility study (Figure 1–9). The statement of need for this stage requires the designer to select the kind of components to be used to make the product or process. The selection process involves studying the trade-offs between alternative types of component hardware and raw materials. It is beyond the scope of this stage to select quantitative things such as dimensional size, critical geometry, color, weight, and so on. For example, in an

Figure 1–9 In the preliminary design stage for a new chemical product, qualitative judgments are made about the relative advantages of design alternatives. (*Courtesy* Union Carbide.)

automobile engine design the preliminary stage would be concerned with the selection of the type of power plant but would not require selection of the exact size of the engine or other details such as the number of bolts used to fasten it to the transmission. Since this stage of design involves the modeling and evaluation of types of elements in general rather than the evaluation of specific sizes and shapes, its output will usually be in the form of sketches, charts, diagrams and reports. Many engineers regard the preliminary design stage as the most interesting phase in the design process because it allows the most opportunity for influencing how the final product or process will look and perform.

Stage 3: Detail stage

The communication output of the preliminary stage will form the basis for a statement of need that requires the complete, detailed specification of the elements recommended for the final product or process. This is called the *detail stage*. The tasks required for the completion of this stage of design will involve making quantitative design selections with respect to size, shape, orientation, color, and so on. The ideas for this stage are relatively easy to create, since most of the major features of the design have already been selected during the preliminary stage. The modeling and testing blocks of the detail stage are often rather complex and involve the use of experiments as well as mathematical analysis. Because exact details of the final result are required, the modeling and evaluation of this stage tend to be extremely detailed and require considerable amounts of scientific thought and analysis. For this reason, the detail stage of design often can be done only by persons highly skilled in a specific technological area. Much of the training provided in engineering and engineering technology programs goes into teaching the professional skills required to perform detail design. The communication output of the detail design stage consists of engineering drawings, parts lists, and assembly drawings. This highly detailed output ordinarily is given to persons involved in the manufacturing and assembly phases of the industry. Once a design has left the detail stage, all decisions about the performance and appearance of the final product have been made and the product or process is ready to be produced and marketed.

Recent technological advances in computer hardware and software have led to increased use of the digital computer for augmentation of the detail stage of design. This exciting new method, known as "Computer-Aided Design" or "CAD," is rapidly growing and becoming one of the most valuable design tools available to the engineer. By using CAD the extreme speed and accuracy of the digital computer can be exploited to assist with the illustration, analysis, and improvement of ideas (Figure 1–10). Some CAD systems are structured so that de-

Figure 1–10 Computer-aided design (CAD) facilitates the development of new products. (*Courtesy* Tektronix, Inc.)

tailed specifications and production information can be transmitted directly to manufacturing equipment without the need for drawings or human operators. This new concept is known as "Computer-Aided Manufacturing" or "CAM." The use of CAD–CAM systems is helping to reduce design costs and production expenses in many industries today. There is every indication that CAD–CAM systems will become increasingly useful in engineering activities of the future.

In most complex engineering systems many different decisions must be specified during the preliminary stage and the detail stage of design. As long as they are independent of one another these decision elements can take place simultaneously, and frequently different persons in a design team will be assigned the responsibilities for these elements. In the diagram of Figure 1–8 this would mean that the design modules for each stage actually contain a number of identical modules that act in parallel.

Sometimes, as a result of the evaluation process in one stage, it becomes obvious that the design would have been greatly improved if a different decision had been made during the previous stage. When this situation arises, it is not uncommon for the flow of the design process to return to the previous stage to capitalize on the suggested improvement. This process is characterized in Figure 1–8 by alternative feedback paths from the "improvement" block of one stage to the "recognition of need" block of the previous stage. Since this feedback process is allowed during the operation of the first three stages of design, it only seems reasonable to allow a similar feedback to occur after the product or process has been manufactured and marketed. When this occurs, the process is called design revision and is listed in Figure 1–8 as the last stage of design.

Stage 4: Revision stage

Once the detailed design has been manufactured and placed in service it may be found desirable to use field experience as a basis for further improvement of the product. This process is called the *revision stage*. Since the final product is available, it is often used as the model for evaluation during the revision stage. The testing and evaluation blocks of this stage may involve "in-factory" experimental tests or may consist of data provided by those persons who use, maintain, or repair the product or process. Automobile manufacturers, for example, maintain extensive testing laboratories and proving grounds to study ways to improve their existing products. Design improvements for increased comfort and safety often come from these studies. Occasionally it may be discovered that some particular part or assembly in an automobile has a rate of failure higher than expected. When such information becomes available, it is extremely useful during the revision process. In the automotive industry such information may form the basis for recalling all vehicles for inspection and replacement of faulty parts if it is believed that the discovery poses a potential danger to the consumer.

The revision stage is, of course, not always used in the design process. Some companies prefer to reinitialize the design process at the feasibility stage rather than to improve an existing product or process.

As an illustration of how the stages of design proceed let us consider the following example.

EXAMPLE 1–1   A major automobile manufacturing company would like to consider production of a new type of personal transportation vehicle.

To perform the feasibility stage of design the company forms a design team made up of two engineers, an economist, a marketing specialist, and one member of the board of directors. The group is charged with assessing the feasibility of a new type of personalized transportation vehicle. The team starts its task by surveying the market demand for such a device. This involves a careful study of the features of the present product line and those of competitors (Figure 1–11). It also involves the modeling and evaluation of the potential market for innovative approaches to transportation. As part of its task, the group also looks into the environmental and social impacts of a new form of transportation. As

Figure 1–11 The copper electric runabout with fiberglass reinforced polyester body. (*Courtesy* Cooper Development Association.)

a result of these efforts, the group concludes that it would be feasible to design and market a new type of vehicle that uses energy more efficiently and causes less environmental damage than do existing vehicles. The group presents its recommendations to the top management of the company in the form of a detailed written report and an oral presentation. Because of the favorable economic possibilities of the new venture, the board of directors of the company decides to move ahead with the project. A new design team is formed to perform the preliminary stage of design. The new team consists of the original two engineers along with a support staff of 30 other engineers.

At its first meeting the preliminary design team subdivides itself into two task forces responsible for the vehicle body and the vehicle power plant. The power plant task force uses the design method to consider alternative forms of powering the new vehicle. It considers diesel engines, gasoline engines, gas turbines, and electric motors. After carefully evaluating the virtues of each of these alternatives, the task group agrees that an electric vehicle is most desirable; the group then turns its attention to the preliminary selection of the source of electric power to be used. As its investigation proceeds, it decides to use rechargeable storage batteries as the energy source. At this point the group adds a battery specialist to help with the other preliminary decisions that must be made. Meanwhile, the body design group begins to deliberate about the type of vehicle to design. The members of the group consider compacts, mid-sized vehicles, and full-sized vehicles. In the final cycle of their deliberations they select a small-sized three-wheeled vehicle. Although they have made decisions about the size, they do not specify the weight or dimensions of the vehicle.

After about three months of intensive work, the two task forces hold their final joint meeting and prepare an extensive written report containing sketches of the proposed vehicle along with specifications for the general equipment associated with the body and power plant of the vehicle. The preliminary design team also makes an organized list of 350 detailed subsystems that must be designed during the next stage of the design process. This part of the written report is used by the company management to select a group of 35 design team leaders who will each be responsible for a specific portion of the detail design work. After the design team leaders have been given a week to study the report of the preliminary design task force, the two groups hold a weeklong series of meetings to clarify details of the design. This is, of course, not the only opportunity for dialogue between the two groups, since a number of the members of the preliminary design team are assigned to the detail design teams as consultants. In addition to the team leader, each detail design team will contain professionals who are specialists in the particular tasks of that team. Each design team also will have support staff in the form of draftsmen, computer-aided design specialists, and laboratory technicians to help with experimental tests and prototype construction. The design teams each contain one representative from the manufacturing division of the company. These persons have been placed on the team to ensure that the final designs are compatible with the manufacturing capabilities of the company.

Each of the design teams uses the design cycle to make critical decisions as to size, weight, and placement of every element in the final design. The body-style design team makes a number of full-sized models of the proposed vehicle in the model shop. The final output of the detail stage of design is an extensive series of engineering drawings of every part to be used in the final product, along with assembly drawings that specify the exact location of each part. These drawings are passed along to the vice president for manufacturing who will see to their implementation. The drawings are also passed along to the vice president for marketing who will immediately begin plans to advertise and market the new vehicle. The result of this overall design process will be the production and sales of a new, compact-sized vehicle that has three wheels and is powered by electric motors using rechargeable storage batteries.

1.4 Design education

Although most of the training in engineering and engineering technology programs is useful in the design process, not all courses in these programs should be classified as design courses. Many engineering and engineering-related courses should be classified as ''analysis'' courses as they provide skills useful for the testing and evaluation phases of the various design stages. A few courses provide training in the design cycle itself. A good example is the course you are taking at the present time using this book as a text.

Design courses at the early undergraduate level tend to be useful for teaching the first two stages of design (the feasibility and preliminary stages), and courses at the upper level tend to be useful for the detail stage of design. The reason for this is that the detail stage of design frequently uses special analysis techniques that require upper level undergraduate training.

Students at the freshman and sophomore levels often find the preliminary design stage to be both interesting and challenging. For many students, courses in design taught at the freshman level represent their first exposure to actual engineering problem solving. Thus, freshman design courses help undergraduate students to explore what the engineering profession is about before selecting a career.

1.5 Summary

Design is the physical and mental activity associated with the creation of products and processes to meet the needs of society. Design is the most creative calling to which an engineer can aspire. To design in a precise, efficient, and organized way is one of the most enjoyable and most rewarding engineering functions. To perform the task well, it is important that the engineer understand the design process and know when and how to apply it. Never in history have the opportunities for engineering design been more challenging. It is in applying their sensitivity to the problems of human need that the talents of today's engineers benefit mankind. In the following chapters we will discuss specific design tools that engineers can use to become better equipped to meet this professional challenge.

EXERCISES

1.1 Name ten human accomplishments of the past 100 years that you regard as most significant. Which of these would you classify as engineering design?

1.2 What is the difference between the process of engineering design and the process of painting a picture?

1.3 An engineering firm is considering the design of a new type of bridge to span a large river. For each stage of design (feasibility stage, preliminary stage, detail stage, and revision stage), what output would you expect?

1.4 Classify the following activities as: FS = feasibility stage, PS = preliminary stage, DS = detail stage, and RS = revision stage:

(a) The environmental impact statement for a nuclear power plant.

(b) Selecting a power source for a new type of hand power tool.

(c) Choosing the color of a new type of concrete building material.

(d) The study of an automobile steering linkage with a view toward recalling all existing linkages in service.

1.5 Explain how a high-school student might use the design method to choose an engineering school to attend.

1.6 State three things that an engineer can do to improve his or her ability for engineering design.

1.7 Make a list of ten problem areas that represent unsolved human needs. How many of these would require engineering design to solve?

1.8 What other professions besides engineering use a design method?

1.9 Where would constraints in the form of laws and code standards fit into the design cycle shown in Figure 1–4?

1.10 What are some of the ways that an engineer can discover the need for new products and processes?

1.11 Within the field of engineering there are a number of different career disciplines such as Chemical Engineering, Civil Engineering, Electrical Engineering, Industrial Engineering, and Mechanical Engineering. How could an engineering student use the design method to choose one of these for a career?

1.12   Using the curriculum catalog of your college or university identify five courses that could be described as courses that teach design or design-related activities. Based on the description and the time in the program that they are offered, classify them according to the stage of design they most likely involve.

1.13   Unlike most analysis problems, design problems have many solutions. To illustrate this fact, perform a preliminary design cycle for the following projects and list at least three possible design choices. (The answers may include products or processes that have already been invented.)

(a) Conceive a device for holding two pieces of lumber together.

(b) Conceive a new type of student transportation system for your campus.

(c) Conceive a new type of study lamp for a college dormitory.

(d) Conceive a device for transmitting printed information from one city to another.

1.14   The practice of design has been around as long as mankind. Imagine that you are in an early agricultural society and are about to invent the wheel. What recognized need does this represent for you? What types of tests would you apply to early prototype models of your device?

1.15   Write a simple, concise statement of need that applies to the following inventions:

(a)   The electric light bulb (Thomas Edison)

(b)   The telephone (Alexander Graham Bell)

(c)   The airplane (The Wright Brothers)

(d)   The cotton gin (Eli Whitney)

(e)   The automobile (Henry Ford)

(f)   The electric automotive starter (Charles Kettering)

(g)   Bifocals (Ben Franklin)

References

ASIMOW, M., *Introduction to Design*. Englewood Cliffs, N.J.: Prentice-Hall, Inc., 1962.

BEAKLEY, G. C., and E. G. CHILTON, *Introduction to Engineering Design and Graphics*. New York: Macmillan Company, 1973.

BEAKLEY, G. C., and H. W. LEACH, *Engineering: An Introduction to a Creative Profession*. New York: Macmillan Company, 1972.

EIDE, A. R., R. D. JENISON, L. H. MASHAW, and L. L. NORTHUP, *Engineering Fundamentals and Problem Solving*. New York: McGraw-Hill Book Co., 1979.

FLETCHER, L. S., and T. E. SHOUP, *Introduction to Engineering with FORTRAN Programming*. Englewood Cliffs, N.J.: Prentice-Hall, Inc., 1978.

GIBSON, J. E., *Introduction to Engineering Design*. New York: Holt, Rinehart, and Winston, 1968.

HILL, P. H., *The Science of Engineering Design*. New York: Holt, Rinehart, and Winston, 1970.

RUBINSTEIN, M. F., *Patterns of Problem Solving*. Englewood Cliffs, N.J.: Prentice-Hall, Inc., 1975.

SHOUP, T. E., *A Practical Guide to Computer Methods for Engineers*. Englewood Cliffs, N.J.: Prentice-Hall, Inc., 1979.

WOODSON, T. T., *Introduction to Engineering Design*. New York: McGraw-Hill Book Co., 1966.

Bicycle windmill. (*Courtesy* Department of Energy.)

Innovation in engineering design

2

Every day a great deal of effort is devoted to improving the quality of our lives. Throughout history, changes in techniques, processes, concepts, and products have altered the condition of human existence. These changes are the result of new ideas and innovations which have provided such benefits as increased comfort, reduced labor, greater convenience, and better health care. The continuing improvement of the quality of our lives is a challenge to the creative and innovative abilities of every engineer.

Innovation and invention have been the backbone of the scientific and technological growth of the United States. The first settlers found it necessary to be innovative in order to survive. As the United States developed, manufactured goods and materials were imported from Europe. The limited availability of goods, their high cost, and the long delivery time required fostered the development in this country of manufacturing techniques and facilities needed to provide for the convenience and comfort of the early Americans.

Shortly after the United States gained independence, a patent act, entitled "An Act to Promote the Progress of the Useful Arts," was approved. As Secretary of State, Thomas Jefferson became the first administrator of the American patent system. He felt, philosophically, that inventions could not be a subject of property, but that society might give an exclusive right to the profits arising from such inventions as a means of encouraging developments and ideas which might benefit society. Jefferson considered inventions a necessity for the development of the country and inspected the latest inventions at every opportunity. His keen interest in invention (Figure 2–1), meticulous notes and drawings, and constant search for better ways to do things contributed significantly to the field of engineering design as we know it today.

Engineering design is by nature innovative. It is essentially the development of a plan for the solution of a real problem, combining creativity with ingenuity and engineering ability. Often engineering design is directed toward the development of new devices or systems to perform some function more efficiently. The success of the engineering design depends upon many factors. Primary among these factors is the creative and innovative component of the design.

In engineering, the terms *innovation* and *creativity* often are used interchangeably. These terms, however, are not synonymous, but actually complement each other in the design process. Innovation generally is defined as the introduction of new methods, ideas, devices, or changes into some process. Innovation is development by modification. It is the introduction of change. Creativity, on the other hand, generally describes the thought process involved when an idea comes into existence or is originated. Creativity is conceiving of change. Harrisberger defines creativity as "a spontaneous, irrational, intuitive process of imagination—a mysterious mixing of previous experiences into patterns and combinations that

Figure 2–1 Mouldboard plow invented by Thomas Jefferson for plowing on hillsides. (*Courtesy* Thomas Jefferson Memorial Foundation.)

are totally new and unique.'' The development of creative ideas is essential to innovation in engineering.

In general, engineering developments are responses to human needs, consumer demands, or the desire for improved performance. Engineers are intrigued by the challenge of developing new designs, particularly when this development increases the value or perceived value of the end product. Value, then, is one of the motivations for engineering design.

Value, of course, is relative, and what is valued by some people in some circumstances may not necessarily be valued by others in different circumstances. For example, water may not be considered particularly valuable unless the reservoir or the well runs dry, and the supply is severely limited. Certainly the value you place on gasoline would soar if your car ran out of gas on a lonely stretch of road in the middle of the night as you headed back from a weekend outing in order to make an 8 A.M. exam. Similarly, the value placed on energy for heating is far greater in the middle of a severe winter than on a warm spring day.

There are many types of value: use value, cost value, time value, exchange value, esteem value, and so on. Each of these types of value may be defined differently by different people at different times. Value, then, is not necessarily related to cost but rather to what someone wants and what he or she is willing to give to have it. The increase in value or perceived value is a stimulus to creativity, particularly in engineering design.

In the first chapter we looked at the overall process of design and discovered that innovation and creativity play a key role in this task. In this chapter we will focus attention on this important ingredient in the design process.

2.1 The creative process

Numerous problems occur every day which demand some form of solution. Problems such as a car that does not start, a calculator with a worn-out battery, a punctured bicycle tire, or a library book listed in the card catalog but which is not in the stacks are reasonably well defined, and their solutions are fairly straightforward. Other problems such as the design of a lunar landing vehicle, the development of a memory chip (Figure 2–2), the development of an anti-collision device for aircraft, or the design of a new type of transportation system are extremely complex, demanding considerable creativity and innovation in their solution. Most of us tend to deal with our daily problems with very little creative effort. However, significant developments or major breakthroughs in any field— from music to medicine to engineering—are the direct result of creative thought.

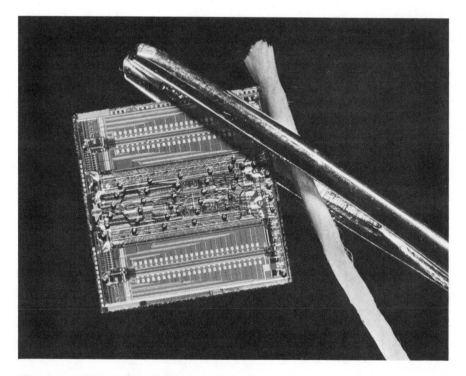

Figure 2–2 IBM's 64,000-bit memory chip in the eye of a needle. (*Courtesy* IBM.)

In engineering, the goal in problem solving should be to devise the best possible solution, providing a wide range of capabilities or increased functions, improved performance, and increased cost-effectiveness. At the same time the solution should be realistic. In engineering as a whole, and particularly in engineering design, it is essential to be more than just creative. The development of new ideas is essential, yet without some form of action the ideas are of little practical value. Thus the engineer is called upon to be both creative and innovative in his or her problem solutions.

Creativity is a mental or intellectual process which may be strengthened through training. It is essential to the development of an engineering design that will enhance the value of the solution to the problem at hand. The creative process is the procedure whereby new ideas are generated.

The creative process generally is considered to include *preparation, incubation, insight,* and *execution.* Figure 2–3 diagrams some of the steps involved in the process. The creative process is, in many respects, a miniature design cycle similar to that described in Chapter 1. It should be noted, however, that unlike the design module, not all of the steps occur in the solution to every problem, and the steps do not have to occur in a specific sequence but may well interact and overlap.

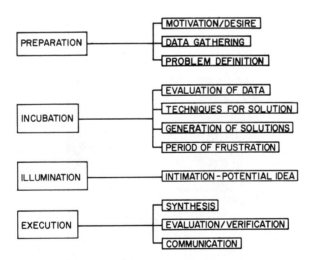

Figure 2–3   Schematic of the creative process.

Preparation.   As an engineer looks toward the challenge of developing a solution to a problem, there must be some *motivation* or desire, some reason for becoming involved in the solution of the problem. The motivation may be personal, it may be humanitarian, it may be job related, or it may be the value perceived or the profit to be derived from the final product.

The engineer's first step is to obtain as much information as possible about the problem. *Data gathering* may involve a library search for written material pertaining to the problem; it may entail a site visit to some location relevant to the problem; or it may involve a thorough discussion with persons familiar with various aspects of the problem. This is the stage in which the techniques needed for handling the information are identified.

Once the engineer is motivated to embark on a project, he or she must *define the problem* precisely in order to orient his or her thinking toward the problem's solution. Specific conditions of the problem must be established and time limitations noted. Careful engineers consider all aspects of a problem as they prepare for a solution. This preparation is essential to an understanding of all ramifications and constraints of the problem and its solution.

Incubation.   The incubation phase of the creative process begins once the problem has been defined and sufficient data and information have been obtained. The next step is the *evaluation* of the information or data obtained in the preparation period (Figure 2–4). This analysis may involve the consideration of various theories, or it may involve rearrangement of the data into forms other than that in which it was obtained. The evaluation of the data allows the engineer to understand thoroughly the available information.

Figure 2–4 Evaluation of preliminary data. (*Courtesy* American Electric Power.)

In addition to evaluating the basic information, the engineer should consider the various possible *techniques for solution* of the problem. The solution may involve numerical methods, computer methods, graphical techniques, or experimental techniques. Often the engineer will evaluate and use several of these techniques in seeking an optimal solution.

From the knowledge gained through the earlier steps of the creative process, several solutions to the problem may seem possible. At this point in the process it would be appropriate to develop as many different or *alternate solutions* as possible, without establishing a priority. Comparison of the potential solutions helps to determine which would be the best possible solution under the existing circumstances.

In trying to evolve a solution, the engineer is often faced with a good deal of *frustration*. No solution may seem appropriate, and no amount of thinking about the alternatives may seem to produce a fruitful idea. The nagging of the mind may create tension, anxiety, or emotional stress. A certain amount of discomfort often can act to stimulate the creative process. This uncomfortable period often is the prelude to inspirational insight.

The incubation process is critical to creative solutions. It is a lonely, individual process involving a period when the mind works over stored information. The engineer is motivated to solve the problem. Much groundwork has been completed, and a great deal of knowledge has been stored in the engineer's mind. Our minds, however, are not computers, and it is not always possible to retrieve or process information on demand. Rather, the mind will work on the problem subconsciously, and at some later time ideas will be transferred back to the conscious mind.

All of us have faced at one time or another a brief lapse of memory. For example, we may try to remember a name, a conversion constant needed on a test, or a telephone number. At the moment we draw a blank, but after some period of time (often a period during which we do not seek a solution) the desired information is retrieved and made available to the conscious mind. A similar phenomenon occurs in the solution to many engineering problems.

Illumination.   The next stage of the creative process is the point at which the conscious mind receives an indirect suggestion or a hint of a suitable solution. This is the moment of *insight;* the potentially good idea has emerged. This phase also is referred to as intimation, illumination, or inspiration.

The period of incubation necessary before the moment of insight can occur will vary widely with the people and the problem involved. A plausible solution may occur after very little thought or, again, the engineer may have to ponder for days, weeks, even years. Since most engineering problems involve groups of engineers, insight may be stimulated through group discussion and interaction. Clearly some problems may be so immense that they seem to defy solution. For most engineering problems, however, a solution can be effected; the insight *will* come.

The illumination phase can be an exciting and pleasant experience. It may occur suddenly when least expected, particularly when no work on the problem is being done—in the middle of the night, while studying for an exam, or during a meal. In essence an idea has "hatched," providing the basis for the solution of the problem. Insight occurs daily, in reading, in puzzles, in games, or in the solution of engineering problems. It is the recognition that something new has emerged.

The engineer must cultivate an awareness of and receptiveness to insights. Discussion of the problem with colleagues with mutual interests is helpful, particularly when the imagination is allowed to play a part in the discussion. From the time the engineer embarks on the problem, he or she should keep in mind the goal being sought. This constant, if subtle, mental pressure will help to bring about creative insight.

Execution.   *Synthesis* is that point in the creative process during which it is necessary to put together the diverse parts of the solution to form a whole. This involves using the knowledge at hand to put together the rudiments of the final solution.

Once the ideas have been synthesized and the solution to the problem has been organized, an *evaluation and verification* of the solution should follow. During this step, the solution must be evaluated to be sure that all conditions of the problem are satisfied and that all aspects of the problem are solved in as thorough a manner as possible within the established time frame (Figure 2–5).

Once the problem and solution have been carefully checked, it is time to "go public" or *communicate* the solution. The scope and limits of the solution should be identified. Special characteristics should be noted and, if necessary, experimental tests should be conducted to demonstrate the viability of the solution.

Clearly all the steps in the creative process will not be required in the creative solution of every problem. The nature of the problem and the solution required will dictate which of the steps could be omitted and which will overlap. In the course of many solutions, this creative process may proceed wholly in the mind and without conscious awareness of the steps.

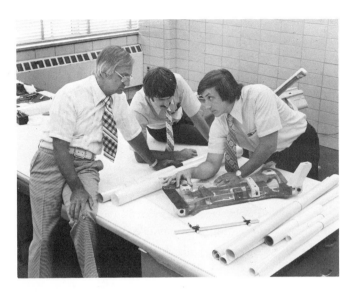

Figure 2–5 Engineers verify that the bracket assembly satisfies all design criteria. (*Courtesy* Chrysler Corporation.)

EXAMPLE 2–1 A new engineer with the XYZ Manufacturing Corporation has been assigned the task of developing a reusable metal fastener to hold together various materials (such as fabrics, plastics, and foils) that the company uses in its manufacturing process. The device must be inexpensive, versatile, and made from materials that are readily available. The fastener may be one piece or may involve more than one piece and must be easy enough to use so that no training or special tools are required.

A creative solution to this design problem might evolve in the following manner:

Preparation. The engineer first considers some appropriate questions:
What are the types of fasteners currently available, their range of application, availability, cost, etc.?
What materials does the manufacturing plant use and which are used the most?
What do other manufacturing companies use?
What types of fastening techniques are advertised in the trade journals?

After considerable preparation, the engineer defines the problem as follows: A device must be made which can hold together a range of materials from light cloth and plastic to medium-weight metallic foils. The device must not be permanently affixed to the material, and use of the device must not tear or make large holes in the substance. It must fasten securely and be small enough to move freely with the material over the mill wheels.

Incubation. The engineer considers zippers, buttons, snaps, hooks and eyes, metallic clips, nuts and bolts with washers, and even bent nails. None of these devices seems to meet all the necessary criteria. The engineer's mind toys with other possible configurations—devising, assessing, discarding, until:

Illumination. Aha! I've got it! The device could be pointed and have some kind of spring and catch to hold it in place.

Execution. The design would involve a piece of wire, sharpened on one end, which is bent so as to create a spring force and a catch or safety shield to hold the point. It would be a one-piece unit, easy to produce in the shop of the manufacturing plant, and quite inexpensive.

Some people call this device the safety pin (see Figure 2–6).

Creativity is not simply inherited. Instead, creativity is based on intellectual capability, experience, knowledge, and the perceived need to develop a new concept or problem solution. Creativity is the exercise of one's imagination. The recognition of needs or value stimulates the creative process. Experience, knowledge, and enthusiasm are equally important in a creative approach to engineering design. The broader the range of experience and education, both formal and informal, the more opportunity there is for creative contribution. Such education should not be limited to technical fields. Most, if not all, of the problems that await the engineer's creative solution have ramifications and effects beyond the purely technical realm. The magnitude of future engineering contributions to the solution of the world's most pressing problems will be determined by the ability of individual engineers to think and work creatively.

UNITED STATES PATENT OFFICE.

WALTER HUNT, OF NEW YORK, N. Y., ASSIGNOR TO WM. RICHARDSON AND JNO. RICHARDSON.

DRESS-PIN.

Specification of Letters Patent No. 6,281, dated April 10, 1849.

To all whom it may concern:

Be it known that I, WALTER HUNT, of the city, county, and State of New York, have invented a new and useful Improvement in the Make or Form of Dress-Pins, of which the following is a faithful and accurate description.

The distinguishing features of this invention consist in the construction of a pin made of one piece of wire or metal combining a spring, and clasp or catch, in which catch, the point of said pin is forced and by its own spring securely retained. They may be made of common pin wire, or of the precious metals.

See Figure 1 in the annexed drawings (which are drawn upon a full scale, and in which the same letters refer to similar parts,) which figure presents a side view of said pin, and in which is shown the three distinct mechanical features, viz: the pin A, the coiled spring B, and the catch D, which is made at the extreme end of the wire bar C, extended from B. Fig. 2 is a similar view of a pin with an elliptical coiled spring, the pin being detached from the catch D and thrown open by the spring B. Fig. 3 gives a top view of the same. Fig. 4 is a top view of the spring made in a flat spiral coil. Fig. 5 is a side view of the same.

Any ornamental design may be attached to the bar C, (See Figs. 6, 7 and 8,) which combined with the advantages of the spring and catch, renders it equally ornamental, and at the same time more secure and durable than any other plan of a clasp pin, heretofore in use, there being no joint to break or pivot to wear or get loose as in other plans. Another great advantages unknown in other plans is found in the perfect convenience of inserting these into the dress, without danger of bending the pin, or wounding the fingers, which renders them equally adapted to either ornamental, common dress, or nursery uses. The same principle is applicable to hair pins.

My claims in the above described invention, for which I desire to secure Letters Patent are confined to the construction of dress-pins, hair-pins, &c., made from one entire piece of wire or metal, (without a joint or hinge, or any additional metal except for ornament,) forming said pin and combining with it in one and the same piece of wire, a coiled or curved spring, and a clasp or catch, constructed substantially as above set forth and described.

WALTER HUNT.

Witnesses:
JOHN M. KNOX,
JNO. R. CHAPIN.

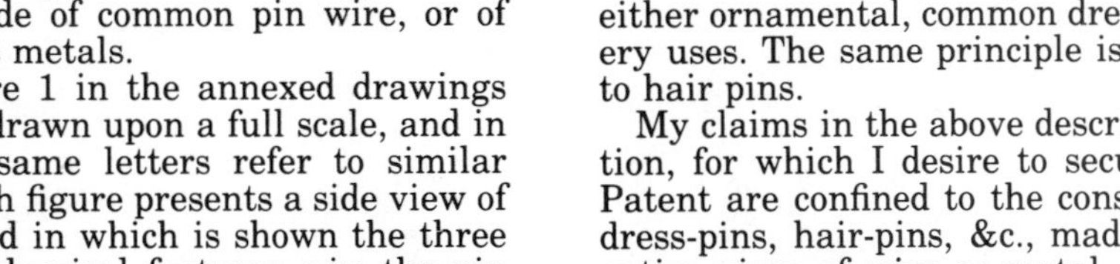

Figure 2–6 Patent issued in 1849 to Walter Hunt for a "pin." (*Courtesy* U.S. Patent Office.)

2.2   Obstacles to creative thought

The history of mankind in general, and of the United States in particular, is a history of innovation and creativity. Benjamin Franklin, Thomas Edison, Alexander Graham Bell, and the Wright brothers (Figure 2–7) are but a few of the most famous American innovators. Thousands of others, from the inventor of the zipper to the developer of freeze-dried foods, have applied their creative thought to the development of new products and the solution of real problems. Yet many potentially creative people find it difficult to make full use of their talents because of certain obstacles to creative thinking.

A number of factors work against the development of creative thinking. Most of these obstructions are internal, resulting from our education, from our life's experiences, and from our individual goals. Obstacles to creative thought may be habitual, cultural, perceptual, or emotional. Our backgrounds, our fam-

Figure 2–7   Wright Brothers' Flyer I making the world's first powered, sustained, and controlled flight in December 1903. (*Courtesy* Smithsonian Institution, National Air and Space Museum.)

ilies, and our peers contribute to the perspective from which we view a problem. Pressure from peers and superiors to think and act realistically dampens spontaneous creativity. In educational institutions, the rigid curricula, the lecture method of classes, and the over-reliance on textbooks and routine assignments further limit creativity. All of these factors influence the manner in which we pursue a creative solution to a problem.

In order to demonstrate some of the difficulties we face in effecting a creative solution to a problem, consider some frequently used examples which demonstrate obstacles to creative thought.

EXAMPLE 2–2   A very hungry bull, 2.5 meters long, 1.3 meters high, and weighing 635 kg is tied very tightly to a strong rope 4 meters long. There is a pile of hay in the bull pen approximately 8 meters away from the bull. Will the bull satisfy his hunger?

At first glance one is tempted to answer, "No." A closer look at the problem, however, reveals that nowhere does the problem mention that the other end of the rope is tied to anything. The answer is that the bull simply walks over to the haypile.

EXAMPLE 2–3   Consider the two lines pictured below. Which one is shorter?

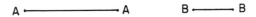

Clearly, when considered in a single dimension the line at the right is shorter. But perhaps the lines exist in two dimensions. If this is so, we must view them from another angle if we are to answer the question accurately.

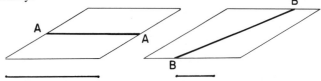

The answer, then, is dependent upon the perspective from which the lines are pictured. It is quite possible that in two dimensions the lines are of equal length or that the one on the left is in fact shorter.

EXAMPLE 2–4   Ten coins are arranged in the following pattern:

Is it possible to move two coins to other positions and end up with two rows of 6 coins?

If all the coins must rest on the table—that is, if the problem must be viewed only in two dimensions—the task is impossible. But no such precondition has been stated. Working in three dimensions it is possible to place two coins at the center (one on top of the other) so that there are indeed 6 coins in each row.

As you may have observed, initial reactions are to assume unnecessary boundary conditions and constraints in the solution of a problem. These initial reactions result from some of the basic obstacles to innovative thought.

Habitual obstacles. All of us have habits: the way we do things, the way we speak, or the way we approach a problem. Often these habits make our lives simpler and more trouble free. But when applied to problem solution, these standard patterns of thought and action may limit our openness to innovation. It is often easier to use familiar tried and true procedures than to develop new and innovative procedures. Our attempt to conform to custom and our reliance on conventional or traditional approaches may well restrict the creative solution of problems.

Cultural obstacles. Cultural obstacles to the creative solution of a problem stem from the environment in which we live. All of us tend to conform to a "proper" pattern of doing things. Local customs and procedures restrict the manner in which we approach the solution of a problem. A student at one university will approach a problem in the way he or she has been taught, while a student at another university may use an entirely different approach. Our educational systems teach us that indulgence in fantasy or daydreaming is a waste of time. Yet daydreaming may lead to unique solutions to unique problems. Engineers are noted for their reliance on reason and logic as a means to the solution of problems (Figure 2–8). Creativity and innovation require, in addition, the application of a certain amount of constructive fantasy.

Perceptual obstacles. Perceptual obstacles also may block the innovative solution of problems. Often it is difficult to visualize remote relationships or to distinguish between cause and effect. Many times we do not use all of our senses for observation, but rather choose the data we want to use. When we use only partial

Figure 2–8 Microprocessors are used to extend an engineer's reasoning in the solution of problems. (*Courtesy* General Electric Company.)

information and fail to investigate thoroughly or to define terms, our view of the problem and its solution may be limited.

Emotional obstacles. As human beings we also confront emotional obstacles or concerns which restrict our innovative thought. Attitudes are often shaped by emotion. Discouragement may lead to apathy about a problem. Those days in which nothing seems to go right often produce an emotional view which influences our attempts to solve a problem. Our egos, our fear of failure in the eyes of others, our distrust of colleagues or subordinates, our suspicion of supervisors or bosses—any of these factors may provide an emotional obstacle to creativity. A desire to succeed quickly may minimize concern for the ramifications or implications of our solution. These attitudes can and do affect the emotional perspective that we have of the problem and its solution.

The foregoing obstacles to innovative thought may be overcome by recognizing them and by conscientiously seeking to reach beyond their limitations. It is important to recognize one's habits, one's cultural background, one's perceptual and emotional perspective, and the manner in which these factors influence problem solutions. In seeking and optimizing a solution, it may be necessary to make and reject numerous ideas and design decisions which are adequate but simply do not lead to the best solution of the problem. It is equally important to be open to new ways and new ideas to foster creativity.

Some additional example problems are given to stimulate the recognition of obstacles to creative thought and to help in their mastery.

EXAMPLE 2–5   As a machinist, you are asked to cut a cylindrical piece of aluminum stock into 8 equal parts using only three cuts of the saw.

As a two-dimensional problem, the task appears impossible. Yet when a third dimension, the thickness of the aluminum stock, is considered the solution is simple. The stock may be cut into 8 equal parts by making two vertical cuts (at right angles to each other) and one horizontal cut.

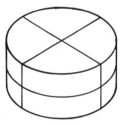

EXAMPLE 2–6   Consider the series of nine dots below.

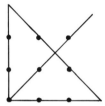

Is it possible to draw four straight lines through all the dots without lifting the pencil or retracing a line?

When looking at the problem, we are apt to assume that the boundaries are confined to the dots. Yet such a restriction is unstated and unnecessary. When this assumed boundary condition is ignored, a solution is easy to accomplish, as shown below:

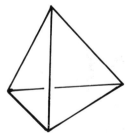

EXAMPLE 2–7   Given six toothpicks, is it possible to make four and only four equilateral triangles by joining the ends of the toothpicks? If we limit our thinking to two dimensions, the problem seems impossible. In three dimensions, however, it is not.

The four faces of the tetrahedron form the four equilateral triangles.

2.3 Spurs to creative thought

How can an engineer be more creative in thought and more innovative in the solution of problems? How can innovative ideas be incorporated in engineering design? What are the steps to the generation of new ideas?

Because many engineering design problems appear routine, a routine method is often employed to effect their solutions. Standard approaches learned in basic courses usually seem much simpler than new and innovative approaches. Consequently, solutions are often generated by minds which seem almost mechanized in their approach and operation.

Creativity can be stimulated by cultivating an open mind and enthusiasm for a new solution or a new perspective on the problem. Attempts must be made to discover new aspects of the problem, to penetrate the problem in order to get to its core. Without a fresh approach, without originality and independence from habits, there will be little opportunity for the generation of creative new ideas.

Many questions can be raised to help spur the generation of new ideas. Can the devices under consideration be put to other uses? Is it possible to adapt an old device to a new situation? Can the existing facility or standard device be modified to accomplish a new task (Figure 2–9)? Will the use of miniaturization contribute to the solution? Is there a substitute approach or technique that can be used? Is it possible to rearrange the components of a device, or even the subcomponents, to obtain better performance? Is it possible to combine, consolidate, or link together standard components to perform a task? These and many other questions may be raised as one seeks a new solution to a problem. The engineer must decide which questions are the most stimulating and useful for the particular problem at hand.

Figure 2–9 Originally designed for pumping water, windmills now are also used to generate electricity. (*Courtesy* Department of Energy.)

Among the many different techniques that may be used to stimulate creative and innovative solutions to problems, several have been found to be particularly effective: brainstorming, matrix techniques, and synectics.

Brainstorming. Brainstorming, involving group participation, is perhaps the most common approach to the generation of ideas (Figure 2–10). Brainstorming is a free-wheeling, everything-goes discussion of methods, concepts, materials, and functions in an open environment. The purpose of brainstorming is to eliminate tunnel vision, to overcome obstacles by focusing attention on a problem from many different perspectives.

A brainstorming group is usually composed of from three to eight persons. Their backgrounds and skills may be diverse; their roles in the solution process may be different; but they share the common objective of reaching an optimum solution to the problem at hand. A brainstorming group may also be called a task force, a project team, a tiger team, a value team, or just a committee. With the right kind of leadership, such groups can accomplish a great deal.

In order to have an effective brainstorming session, some ground rules must be agreed upon by the group. There must be an open environment. Participants must understand that the objective of the session is to create ideas, and they must not become involved in personal bickering. The session should be nonjudgmental. Each participant should be given an opportunity to be heard, and all ideas should be acknowledged and recorded for future reference. These ideas should be judged and evaluated after the brainstorming session is completed. If there are areas in which the group has little or no expertise, specialists may be called in to provide the needed information. In brainstorming, the ideas generated by one person may serve as a stimulus to another, producing a snowball or synergistic effect. The result may be the generation of solutions which combine the input of several members of the group.

Figure 2–10 A brainstorming session. (*Courtesy* General Electric Company.)

EXAMPLE 2–8   The Board of Directors of the Fantastic Foods For Tomorrow Corporation (FFFTC) reviews projected earnings for the coming year. They express concern about the marginal profit expected in the Fantasto Flakes Department, due to cost increases in packaging materials and market competition which keeps the shelf price low.

The chief engineer for the Fantasto Flakes Department is told to develop a more efficient, less costly way to package Fantasto Flakes. She decides to form a task group to investigate the problem. Several design engineers are brought together with specialists from other divisions of the company, including experts in chemistry, the environment, materials, nutrition, marketing, manufacturing, and customer relations. All the participants in the task group are concerned about the problem, and all have a vested interest in the company.

An agenda for the first meeting is established, and a tentative time is set for completing each item on the agenda. At the meeting, the seriousness of the problem is explained, and the current techniques for packaging Fantasto Flakes are described. To begin the group discussion, the chief engineer has a list of "thought starters" which are used to stimulate the group. The list includes such questions as:

—What additional aspects of packaging lines can be automated?

—Can microprocessors be used to regulate the packaging?

—Would packaging in plastic containers be less expensive than cardboard boxes?

—Can energy be conserved in the packaging process?

—Would a rearrangement of the packaging line reduce personnel time?

—Can recycled materials be used in packaging?

During the free-wheeling brainstorming session, the problem is attacked from every conceivable perspective. Each participant offers ideas based on his or her own expertise. All significant comments and suggestions are written down for possible future reference and evaluation. As idea builds on idea, a number of possible solutions evolve. After the meeting, the chief engineer investigates and evaluates the various potential solutions. Once the evaluation is complete, the chief engineer will reconvene the task group to consider the evaluation and to offer additional suggestions.

It is an interesting characteristic of brainstorming sessions that some ideas which seem ridiculous or absurd at first may later turn out to be extremely useful. For this reason it is important to remember to refrain from judging ideas until after the brainstorming session is over. Experience has shown that the larger the number of ideas generated, the higher will be the probability of finding an outstanding solution.

Brainstorming, then, provides an opportunity to capitalize on the ideas of others, compounding the benefits derived from the response to each idea presented. It is, perhaps, the most often used technique for the generation of new ideas.

Matrix techniques. A matrix is an array of elements which can be manipulated according to certain rules. The matrix approach discussed here is one in which ideas can be developed from various groupings of information. There are several types of matrix approaches to creativity, including attribute listing, morphological chart analysis, input-output techniques, and function analysis. Although each of these techniques may be useful in specific situations, three are more commonly used in engineering.

The technique of *attribute listing* involves the listing of the characteristics, qualities, or properties of a material, device, or problem area. This technique may be used to identify the essential characteristics of the problem itself or those of the desired solution. The objective of the attribute list is to stimulate our thinking process, to generate new combinations or new characteristics which might better solve the problem at hand. The process of attribute listing helps to clarify the essential characteristics of the problem and its desired solution.

EXAMPLE 2–9   If you were asked to design a new jogging shoe or a new library study carrel, you might list some attributes of the product in order to clarify the basic requirements for design:

Jogging Shoe	*Library Carrel*
traction	workspace
durability	lighting
air circulation	noise isolation
weight	construction
flexibility	durability
comfort	shelf space
size	privacy

Attribute listing is an individual activity. Different people will interpret a problem differently and will come up with diverse lists of attributes. Whatever the specific content of the resulting list of attributes, the process through which it is developed focuses the mind on the basic problem and stimulates creative solutions.

Checklisting is a technique used by many engineers to identify different ways of viewing a problem. The development of a list of questions concerning the general area of the problem is perhaps the most common checklisting approach. Each question on the list may bring to mind other questions which might be raised. Once the checklist is made, it is helpful to ascertain which of the questions are of primary importance and which are of secondary importance to the solution of the problem.

The purpose of the checklist questions is to point up other ways of looking at a problem and thus to stimulate creative thinking. By asking ourselves thought provoking questions, we open the opportunity to investigate side effects of the problem. In essence we are providing conscious guidance to the formulation and solution of the problem. The more questions that are asked and the more exploration of unusual characteristics that is undertaken, the more likely it is that new and innovative ideas may occur.

EXAMPLE 2–10   One of the automobile manufacturing companies has asked your firm to design and develop an improved air pollution control (APC) device for lightweight vehicles. As senior design engineer, you are responsible for initiating the design project, and you develop the following list of questions for your design staff:

—Could the currently used APC unit be modified for this application?

—Should the new APC unit be larger/stronger than the present unit?

—Could the present APC unit be miniaturized for this purpose?

—Is it possible to rearrange the components of the present APC unit to be more effective?

—Could new technology be combined with the present unit to meet the needs?

—Is there new technology which could be used to design a totally new APC unit?

—Could the newly designed APC unit be used for other purposes?

The morphological chart technique, or *morphological analysis,* is another means of stimulating the imagination. It involves the development of a list of independent parameters or characteristics associated with the problem, along with several design alternatives for each parameter. A typical morphological chart is diagrammed in Figure 2–11.

Independent Parameters	Design Alternatives			
W	W_1	W_2	W_3	W_4
X	X_1	X_2	X_3	
Y	Y_1	Y_2	Y_3	Y_4	Y_5
Z	Z_1	Z_2	Z_3	Z_4

Figure 2–11 Schematic of a morphological chart.

The chart permits a comparison of different characteristics which might not otherwise be associated. For example, in Figure 2–11 alternatives $W_2 X_3 Y_1 Z_4$ might be combined to form a very interesting solution—or they might prove to be an impossibility. Such an approach will enable the investigation of many different combinations of variables which checklisting would not permit.

EXAMPLE 2–11   As an engineering student, you have been asked to design a drawing table for a graphics laboratory. After evaluating all of the pertinent information, you develop the following morphological chart.

Independent Parameters	Design Alternatives					
shape	round	square	rectangular	triangular		oval
material	aluminum		wood	plastic		steel
height	0.8m	0.9m	1.0m	1.1m	1.2m	1.3m
support	pedestal	2 legs	3 legs	4 legs	6 legs	box
angle	5°	10°	15°	20°	25°	30°	variable
storage	1 drawer	2 drawers	shelf	cabinet	lift-up lid

It is evident that there are many design choices. The table most commonly used is noted by the dashed line. Yet the only clear requirement is a flat surface. What creative alternatives can you devise?

Matrix techniques may be used individually or in combination to provide new ways of evaluating information pertinent to a problem. Each of the techniques is appropriate for a particular group of problems. Some engineers feel more comfortable using one approach, while others prefer to use different approaches for different problems. The use of matrix techniques in creative thinking is a function of both the problem and the individual involved in seeking a new solution. Judicious use of matrix techniques can result in unique design alternatives (see Figure 2–12).

Synectics.   The term *synectics* comes from the Greek and is defined as the combination of a number of different and apparently unrelated elements. It is an operational theory for creativity sometimes called the Gordon technique after its inventor, W. J. Gordon. The synectics process provides insight into the concepts which underlie the problem at hand and is particularly useful in seeking new products or new uses for existing products. The technique deals primarily with the use of analogies—personal analogies, direct analogies, symbolic analogies, and fantasy analogies. The first three are most appropriate for use in engineering.

In using the *personal analogy,* the individual personally identifies with specific characteristics or elements of the problem. The problem is investigated or viewed from the perspective of that particular element. As an example, consider the design of a new device. Imagine that you are one of the components of the

Figure 2–12 A joined-wing aircraft designed for agricultural applications. (*Courtesy* NASA.)

device that you are trying to design and ask questions of yourself about how the other parts should move or what the arrangement of components should look like.

The *direct analogy* involves the actual comparison of characteristics from totally different systems. Fruitful comparisons have been found between nature and many scientific and engineering developments. For example, the comparison of a tree with a complex energy distribution system can often give useful information on how to organize the system. *Symbolic analogies* are more abstract and use impersonal images to assist in the solution. Characteristics of the problem are assigned symbols or images, and the symbols are combined in different manners to form aesthetically acceptable patterns.

Synectics also may involve an investigation of the basic characteristics of a problem area. For example, in considering an analysis of energy distribution in the city, it would be appropriate to investigate the basic forms of energy used and the techniques employed for energy conversion. In designing a device for minimizing the use of fuel, it would be appropriate to investigate the types of materials and the kinds of fuels which might be used.

2.4 Summary

Creative thinking is essential to innovative engineering. Creativity can be developed through use of the creative process, which provides a systematic approach to the solution of problems. Our standard patterns of behavior, our cultural and educational environments, and our emotional and perceptual attitudes may pose obstacles to creative thinking. These obstacles may be overcome through conscious effort and use of techniques specifically devised to stimulate creative ideas.

Creative and innovative engineers are essential to a progressive, technologically oriented society. Being creative is demanding; it takes courage and conviction and a conscious desire to develop new techniques and new solutions. The problems which face humanity today are immense. Significant contributions to the solution of these problems will be made by those engineers who cultivate and use their innate creative abilities.

EXERCISES *Group discussion topics*

2.1 Public transportation is a major problem in most cities. What are some improvements that could realistically be implemented in the next 5 years?

2.2 In order to provide barrier-free access for handicapped students, what changes could be made in your university?

2.3   In view of the current energy dilemma, what are ten ways that your school or college could reduce energy consumption and save on operating expenses?

2.4   Crutches allow the body to swing from the arms, and the design of such devices has been basically unchanged for centuries. What innovations would you propose to make crutches more comfortable to use?

2.5   A new land development is being planned, and the developer would like attractive yards. What are the alternatives to grass?

2.6   Noise in dormitories is an impediment to study. What can be done to ease the problem?

2.7   What factors must be considered in developing a wheelchair to climb stairs?

2.8   How should ethics and professionalism be considered in developing creative solutions to problems?

2.9   Parking for students is a serious problem on most university campuses. Is there a better way to handle parking and transportation on campuses?

2.10   To make introductory engineering courses more interesting, what changes should be made in the course format and material coverage?

2.11   Conduct a brainstorming session to suggest possible uses for the following:

(a) pull-tab rings from beverage cans

(b) used toothbrushes

(c) used automobile tires

(d) used ball-point pens

(e) plastic beverage bottles

2.12   List the attributes of the following designs:

(a) tennis racket

(b) wooden lead pencil

(c) comb

(d) garden shovel

(e) paper plate

2.13   What obstacles to creative thought are present in the following design tasks?

(a) the design of a new type of car

(b) the creation of a new type of sandwich

(c) the design of a new type of chair

2.14 What design innovations might be possible if these obstacles are removed?

2.15 What similarities do the following design objects have in common and what uses for design improvement could be made through a careful consideration of these similarities?

(a) a dinner plate and a phonograph record

(b) a piano and a typewriter

(c) a rocking chair and a swing

(d) a washing machine and a hot tub

(e) a telephone and a clock radio

Individual exercises

2.16 Fuel consumption in an automobile is proportional to the weight of the vehicle. What changes would you propose to automobile manufacturers that would lighten vehicles but still maintain structural integrity?

2.17 List three major problems at your institution and outline how you would approach a solution.

2.18 List three ways in which a minicomputer could be used in your home or apartment to facilitate daily chores.

2.19 The use of energy is of great concern. What specific changes in your daily life-style could be made in order to reduce your energy consumption?

2.20 Suppose that every home had a computer terminal. List ten ways in which the terminal could be used to change our life-styles.

2.21 List five ways to use discarded aluminum cans, other than in the recycling process.

2.22 Can you identify three things you have accomplished that have been significant factors in your professional growth?

2.23 List the professional societies you plan to join upon entry into the engineering profession and your reasons for selecting these particular societies.

2.24 A motion picture company is developing a film on interplanetary travel. You are asked to design the inhabitants for planet Heliospirus. Devise a morphological chart to aid in the design of the inhabitants. Assemble several possible alternatives.

2.25   What are your major obstacles to creative thought?

2.26   List the major reasons for selecting your discipline in engineering.

2.27   List five ways in which professional engineers could make a contribution through service organizations in the community in which they live.

2.28   List three ways to encourage students to use bicycles.

2.29   What kind of useful work could be performed by domestic dogs and cats?

REFERENCES

ARMSTRONG, F. A., *Idea Tracking*. New York: Criterion Books, 1960.

BEAKLEY, G. C., and H. W. LEACH, *Engineering: An Introduction to a Creative Profession,* 3rd ed. New York: The Macmillan Company, 1977.

BEAKLEY, G. C., and E. G. CHILTON, *Introduction to Engineering Design and Graphics*. New York: The Macmillan Company, 1973.

DAUW, D. C., and A. J. FREDIAN, *Creativity and Innovation in Organization— Applications and Exercises,* 3rd ed. Dubuque, Iowa: Kendall/Hunt Publishing Company, 1976.

GIBSON, J. E., *Introduction to Engineering Design*. New York: Holt, Rinehart and Winston, Inc., 1968.

GORDON, W. J. J., *Synectics*. New York: Harper & Row, 1961.

HAEFELE, J. W., *Creativity and Innovation*. New York: Reinhold Publishing Company, 1962.

HARRISBERGER, L., *Engineersmanship—A Philosophy of Design*. Belmont, California: Brooks/Cole Publishing Company, 1966.

KRICK, E. V., *An Introduction to Engineering: Methods, Concepts, and Issues*. New York: John Wiley and Sons, Inc., 1976.

KNELLER, G. F., *The Art and Science of Creativity*. New York: Holt, Rinehart and Winston, Inc., 1965.

LUZADDER, W. J., *Innovative Design with an Introduction to Design Graphics*. Englewood Cliffs, N.J.: Prentice-Hall, Inc., 1975.

MALONE, D., *Jefferson and His Time*, vol. 2, *Jefferson and the Rights of Man*. Boston: Little, Brown and Company, 1951.

MARTIN, E. T., *Thomas Jefferson: Scientist*. New York: Henry Schuman, Inc., 1952.

OSBORN, A. F., *Applied Imagination—Principles and Procedures of Creative Thinking*. New York: Charles Scribner's Sons, 1953.

OSBORN, A., *Your Creative Power—How to Use Imagination*. New York: Charles Scribner's Sons, 1948.

PETERSON, M. D., *Thomas Jefferson and the New Nation—A Biography*. New York: Oxford University Press, 1970.

RUBINSTEIN, M. F., *Patterns of Problem Solving*. Englewood Cliffs, N.J.: Prentice-Hall, Inc., 1975.

SMITH, P., ed., *Creativity—An Examination of the Creative Process*. New York: Hastings House, Publishers, 1959.

SMITH, R. J., *Engineering as a Career*, 3rd ed. New York: McGraw-Hill Book Company, Inc., 1969.

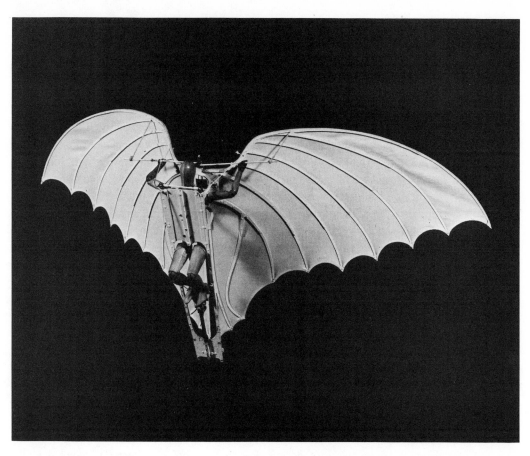

A model of Leonardo da Vinci's flying machine constructed from his sketches. (*Courtesy* IBM.)

Model building

3

Once a design idea has been created in the mind of its inventor, it becomes necessary to test and evaluate the idea. The creation step of design is connected to the testing step of design by means of a model. In the engineering design process, the term "model" is used to describe any physical or mental abstraction of an idea. Sometimes an engineering model will be physical and will physically resemble the idea being considered. For example, a prototype model of a new body style for a car will bear a strong visual resemblance to the new design. Sometimes an engineering model will be abstract and will bear no physical resemblance to the design being considered. Such abstract models are often used to predict performance rather than appearance. A good example of an abstract model would be a mathematical formula used to predict the fuel consumption of a vehicle based on its weight, speed, and number of kilometers driven. Even though the formula itself can be seen, it bears no physical resemblance to the vehicle, the fuel, nor the roadway. Sometimes an engineering model may be entirely mental. All manipulation and evaluation of this type of model will take place in the mind of the inventor. Although every model starts as a mental image, it is not uncommon for a number of different types of models to be used to test and evaluate an engineering idea.

Creating a good model of an engineering idea can often be as exciting and as challenging as the creation of the design itself. The ability to formulate and to manipulate engineering models is an extremely important talent for every designer to cultivate. Experience has shown that the more different types of models an engineer is able to understand and to use, the better he or she will be at performing the professional tasks of design. For that reason this chapter is devoted to the topic of engineering models. We will first discuss the important characteristics of engineering models in design. We will then look at the spectrum of available types of models for engineering practice. Finally we will look at the role of the engineering model in the design process.

3.1 The characteristics of engineering models

Regardless of whether they are abstract or real, large or small, all good engineering models should exhibit certain common characteristics. The degree to which a model is capable of exhibiting these characteristics will determine its overall usefulness in performing its intended function (Figure 3–1).

A model must be functional

To be effective, an engineering model must be susceptible to either physical or abstract manipulation in order to exercise and test the key features of the idea

Figure 3–1 Engineers study the performance of a new typewriter design by means of a breadboard model that duplicates the functional behavior of the proposed design. (*Courtesy* IBM.)

it represents. For example, a model for testing the motion of a new type of folding chair must be capable of being folded like the chair. Thus a scale model constructed of toothpicks and beverage straws might serve as an excellent model for the illustration of motion. It should be kept in mind, however, that this same representation might be unsuitable for testing the weight capacity of the chair. Clearly the toothpick structure cannot model the structural behavior of the actual chair. A device or mental image that cannot be tested in a meaningful way is useless as a model no matter how interesting or attractive it may otherwise appear.

A model must be economically practical

Some designers regard the completed design as the ultimate model. While it is true that there is no better representation for an engineering design than the final product itself, in reality this type of model is seldom practical. This is because the cost of the material and manpower to construct a project can be large. In most circumstances it is not practical to build a completed project from every idea that is generated during the idea creation step of design. Thus a model is usually a representation of the design idea that is less expensive to build and modify than the final product itself. An inexpensive model usually can be modified easily and can even be discarded without significant economic waste. Since modifications generally will improve the design, an inexpensive model will often facilitate design improvement. In addition, an inexpensive model will contribute less to the overall cost of the design process than would a more expensive model. In the case of the toothpick model used in the design of the folding chair, several hundred different models could be built and tested for less material cost than would be required for a single unit of the final product. Choosing an appropriate model often involves a compromise between function and cost. An engineering model must be complex enough to duplicate the essential elements of an engineering idea but must also be simple enough to be economical (Figure 3–2).

A model must be timesaving

Another reason for not using the actual product as its own model is that the design and construction time involved may be extensive. In order to evaluate and improve an idea in a minimum amount of time, the design engineer needs models that allow rapid alteration and improvement. The more ideas that an engineer creates, the higher will be the probability of discovering an outstanding solution to the stated need. Thus a functional model that takes little time to construct and

Figure 3–2 The engine simulator shown in this photograph is an economical model because it can be reused for many different design projects. *(Courtesy* Woodward Governor Co.)

little time to modify is an efficient tool for design because it allows the user to evaluate a number of alternative ideas in a short time.

Since engineering salaries are rather high, the engineer should view time-saving models as a way to provide the most engineering output for a given expenditure of effort and dollar resources. As the design process proceeds, the time required to construct an adequate model tends to increase because of the need for more and more detail. Thus in the early stages of design, a crude, quickly constructed model will often suffice. Later, during the detailed design stages, the model will usually be prepared with considerable care and precision. For example, during the preliminary design of a bridge, the designer might use a simple sketch as a model for evaluation. The sketch would likely show only the gross features necessary to complete the preliminary design stage. Also, the sketch would usually be quickly drawn. Later, during the detail design stage, the location of every truss, rivet and gusset plate must be illustrated in exact geometric form. The model for this stage will likely be an extensive engineering drawing that may require considerable time to construct.

A model must be succinct

The complexity of a model need only be sufficient to analyze the important features of an idea. The best possible model is one in which the important decision parameters are presented in vivid detail and the unimportant parameters are minimized. Such a succinct model facilitates treatment of the important elements of the design, free from unnecessary complications. For example, in the design of a particular threaded fastener to join two parts of a machine, a simple pencil sketch is an excellent model because it emphasizes the importance of size and shape and minimizes less important factors such as color and surface finish. In the previous section we noted that the time required to prepare a model depends in part on the stage of design in which it is to be used. Thus, the use of a simple model that is free from unnecessary details is especially important during the detail stage of design.

A model must have a functional size

Since one of the primary roles of a model is to facilitate a prediction of the quality of a given design idea, it is important to have a model that is of a manageable size. When a physical model is used, it will usually be of a size suitable for tabletop manipulation. If the actual design is quite large, as would be the case for an airplane wing or a reservoir, the model will usually be scaled to a fraction of the size of the final product. If the actual size of the design is quite small, as would be the case for many electronic components (Figure 3–3), the model will usually be constructed many times larger than the actual size. A full-scale model is useful when evaluating appearance, perspective, or the interrelationships among existing elements. For example, a full scale mock-up of an airplane cockpit allows the designer to see that all controls are within easy reach of the pilot.

3.2   Types of engineering models

As the process of design takes place, an engineer will often make use of a variety of different models. An engineer who is proficient in the use of many different

Figure 3–3   A circuit board for applied digital data system with numerous miniaturized electronic components. (*Courtesy* General Electric.)

models is like a mechanic with many tools or a writer with a large vocabulary. The more types of models that an engineer can use, the better equipped he or she will be to perform the tasks associated with the design process. We will now look at some of the frequently used engineering models.

Mental models

Perhaps the most frequently used engineering model is a mental image of the design idea. This model has several significant advantages. Because it calls upon the resources of the human mind and can be constructed, modified, and improved with incredible speed, mental models are extremely useful in the design process. The ability to create mental images of a design is a talent present in varying degrees in all persons. Although this ability can be expanded through experience and practice, there is a practical limit on the amount of detail that the human mind can visualize at any given time. That is why engineers often turn to other types of models to supplement their mental ability.

Word models

One of the most frequently used devices to supplement a mental model is the use of a verbal or written description of the design idea. Word models in engineering often include considerable quantitative and qualitative information in order to provide a clear representation of the idea they describe. A common word model in engineering is the parts list that accompanies a set of engineering drawings. The parts list serves a vital role not only as a model to allow organization and manipulation of a design but also as a way to communicate the details of the design to the next phase of the engineering process. In the field of engineering design, certain types of word models are used so frequently that a special vocabulary of abbreviations is available for the designer to use. A good example is found in the word models used to specify threaded fasteners. Chapter 8 treats this topic in considerable depth.

Word models are often used in combination with visual models to provide a greater potential for manipulation and communication.

Sketches and drawings

Through simple sketches, an engineer can gain considerable insight into the size, relative proportions, and orientation of the various elements of an idea. Since inexpensive paper and pencil sketches are easy to make, they are frequently used to supplement the mental process. The use of sketches to communicate an idea dates back to the time when simple sketches on the walls of caves were used as a means of visual communication. Leonardo da Vinci is well-known for his design sketches, many of which were quite advanced for the age in which they were produced (Figure 3–4). Sketching is usually done freehand, with size and proportions often arbitrarily selected to be pleasing to the eye. As an aid to sketching, engineers often make use of special pads that have vertical and horizontal ruled lines to help with the selection of proportions. A carefully made sketch of a simple part is often all that is needed to communicate the design to the manufacturing stage. An engineering sketch made in pencil can be modified and improved with a minimum of effort (Figure 3–5).

A more advanced form of sketching is found in the engineering drawing. In a drawing, the exact proportions and size of the object being represented are maintained, and special equipment is used to ensure that lines are straight and parallel. Quite frequently the sketches of an engineer will form the basis for the generation of detailed engineering drawings. Such a drawing not only contains a graphical picture of a design but also contains dimension specifications and explanations of key features relevant to its construction.

Unlike design sketches, good quality engineering drawings take time to make and involve a number of rules and conventions that have become established standards of practice. Although today's engineer will usually have a skilled draftsperson to convert the sketches and ideas into drawings, it is nevertheless important that the engineer have a good working knowledge of the standards of practice in the generation and interpretation of engineering drawings (Figure 3–6). It is for

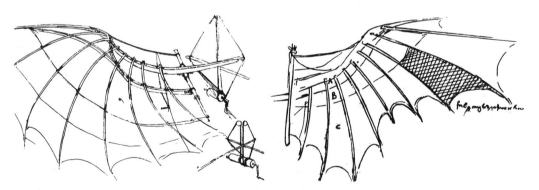

Figure 3–4 Leonardo da Vinci's sketches of a flying machine. (*Codex Atlanticus, courtesy* Alderman Library, University of Virginia.)

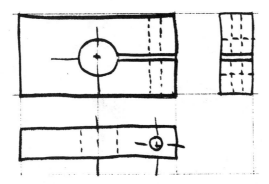

Figure 3–5 A sketch such as this allows the rapid manipulation and communication of geometric details that would otherwise be difficult to visualize.

this reason that considerable attention is given in this book to the topic of engineering graphics.

One of the inherent disadvantages of sketches and drawings is that they are two-dimensional while the objects they describe are frequently three-dimensional. To overcome this shortcoming, multiple views of an object are often used. An alternative approach involves the use of pictorial drawings such as isometric and

Figure 3–6 Engineers use a drawing to discuss the details of a construction project. (*Courtesy* Houston Lighting and Power Company.)

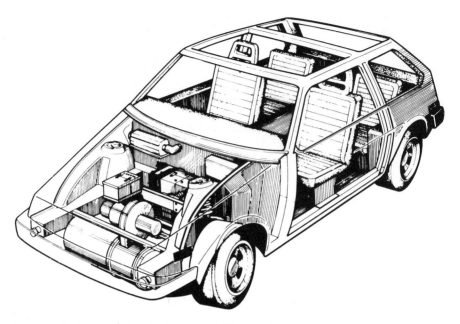

Figure 3–7 The pictorial drawing of this flywheel car provides three-dimensional insight into the placement of different components within the drive system. (*Courtesy Mechanical Engineering.*)

perspective. Such special drawings provide graphical models that enhance insight into the spatial relationships among the various details of a design (Figure 3–7).

Diagrams

Diagrams are somewhat like sketches and drawings, the primary difference being that a diagram may not bear a physical resemblance to the design it is used to model. This basic characteristic allows engineering diagrams to be used to model abstract concepts. For example, the logic flow process that an engineer uses to solve a problem may be modeled by a logic flow diagram (Figure 3–8). In this diagram, the steps in the process are modeled by symbolic blocks. In Chapter 1 the design process itself was described in terms of a functional diagram.

A diagram allows the designer to manipulate an abstract idea so that the interactions among elements can be clearly understood. A frequently used model in engineering practice is the free body diagram (Figure 3–9). This type of diagram allows forces to be represented as vectors. The free body diagram is actually more than a diagram since it also involves a sketch or drawing in addition to the force vectors. The free body diagram often forms a basis for the preparation of a mathematical model using the physical laws of motion.

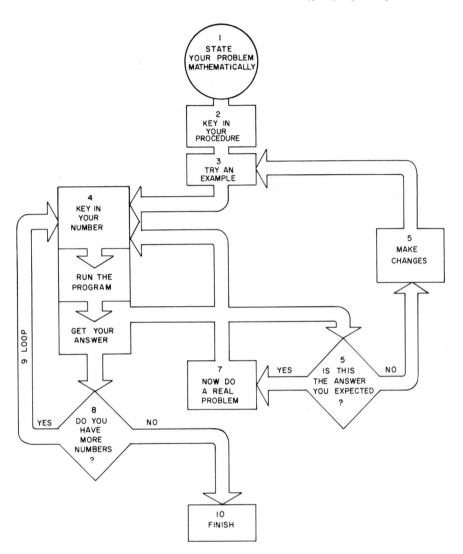

Figure 3–8 A logic diagram of the ten steps necessary to solve a problem on a programmable calculator. (*Courtesy* Texas Instruments.)

Mathematical model

Most engineering problems involve principles of chemistry and physics. These principles are usually described using mathematical formulas. Thus many engineering models are mathematical. One of the strong advantages of a mathematical model is its ability to provide quantitative information as its output. Ex-

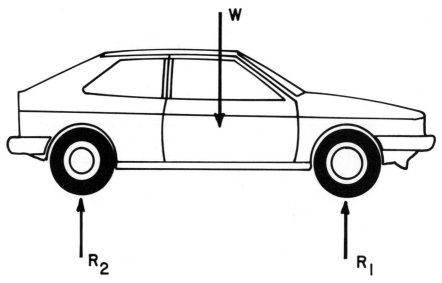

Figure 3–9 The free body diagram is a useful tool to allow manipulation of forces as vectors.

amples of frequently used mathematical models in engineering include Newton's laws of motion, equations of chemical equilibrium, Kirchhoff's current and voltage laws, and energy balances. To illustrate a mathematical model that might be used in engineering practice, let us consider a simple example.

EXAMPLE 3–1   A manufacturer of swimming pool accessories is considering the design of a new type of inexpensive diving board, as shown below.

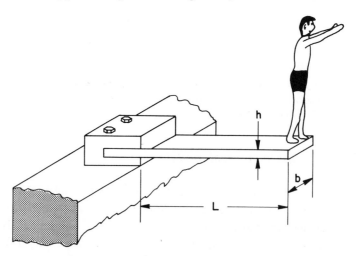

As a part of the evaluation of this new design, the chief engineer needs to know the static deflection of the free end of the diving board when an 80 kg person is standing on the end. To treat this problem, a simplified diagram can be drawn:

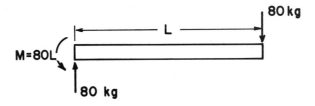

This model presents the diving board as a simple cantilever beam. For such a beam the Euler-Bernouli equation will provide the end deflection by the following formula:

$$\delta = \frac{PL^3}{3EI}$$

where: $E =$ the elastic modulus
$L =$ the board length
$P =$ the end load
$I =$ the moment of inertia for the cross section

$$\left(I = \frac{bh^3}{12}\right)$$

Using this formula and the following data: $E = 6.9 \times 10^5$ N/m², $L = 3.0$ m, $P = 80$ kg, $b = 0.4$ m, and $h = 0.03$ m, the engineer predicts an end deflection of 0.128 m. If this deflection is unsatisfactory, the designer could use the mathematical model as a basis for changing the deflection. For example, if more deflection were desired, the value of L could be increased or the values of b and h could be decreased. It should be noted that the formula shows that the design is much more sensitive to changes in the thickness "h" than it is to changes in the board width "b."

The design is, of course, not only dependent on the end deflection. It is also important to check the strength of the diving board to make sure that it does not break during use. To do this the engineer might make use of the formula:

$$\text{Stress} = \frac{6PL}{bh^2}$$

which predicts the maximum stress level at the fixed end of the diving board where failure is most likely to occur. Using the previous design values and this formula, the engineer predicts the stress level to be 4×10^6 N/m^2. A careful comparison of this value with the maximum allowable stress for the material being used will help the designer to decide if the design is strong enough. The actual comparison would, of course, take into account the following additional factors:

1. Is it possible to have a higher stress due to impact?
2. Is it possible to have a heavier person or more than one person on the board at a single time?
3. Will the material properties change with prolonged use and exposure to the pool environment?

Because of its ability to handle numerical manipulations accurately and rapidly, the digital computer is often used to exercise a complex mathematical engineering model (Figure 3–10). Recent advances in interactive computer graphics have produced computer systems capable of assisting the designer in the creation, manipulation, and improvement of design ideas. These special computer systems actually combine graphical models with mathematical models to enhance the modeling process (Figure 3–11).

Physical models

Because they provide considerable insight into the interrelationships between component parts, physical models are frequently used in engineering design. One commonly used physical model is a scale model of a factory or processing plant (Figure 3–12). Using such a model, the engineer can move machines and equipment around and can thus find the most efficient layout.

Another commonly used physical model is the breadboard model which is a temporary working model of the final design (Figure 3–13). In the breadboard model the components are temporarily fastened in place to observe their operation. Breadboard modeling is frequently done in the design of electronic circuits, process control systems, hydraulic systems, and pneumatic systems. Once the

Figure 3–10   Computer terminals serve as a valuable link with digital computers. (*Courtesy* IBM.)

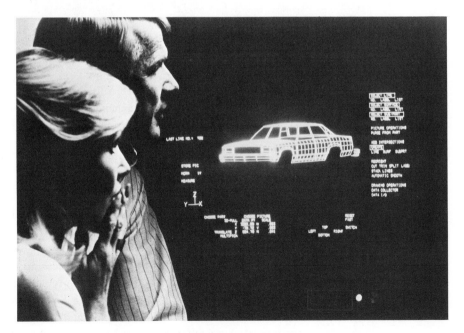

Figure 3–11   Automobiles are designed using interactive computer graphics. (*Courtesy* Ford Motor Company.)

Figure 3–12　This model of a nuclear generating facility provides a spatial view of the equipment. (*Courtesy* Houston Lighting & Power Company.)

breadboard model performs in an optimum manner, the engineer can have a more permanent version of the design constructed using the elements selected by the design process.

Analog models

In an analog model, the physical quantity being investigated is represented by some other quantity that is more easily manipulated. Thus, for instance, the flow of an explosive chemical may be simulated by the flow of electricity in an electronic circuit (Figure 3–14), and the behavior of a spacecraft during reentry may be modeled by an experiment in a water tank. The reason that the analog principle can be used to solve engineering problems is that a diversity of physical phenomena behave in a similar manner. For example, there is a great deal of similarity among the flow of electricity, the flow of heat, and the flow of fluids. Thus an electronic circuit can be made that will have a current flow that duplicates the flow of a fluid in a complex hydraulic system. It is usually easier, less expensive,

Figure 3–13 The breadboard model allows for actual testing of a proposed system before that system is finalized. (*Courtesy* IBM.)

and less hazardous to experiment with an analog model than with an actual hydraulic circuit. The analog modeling concept forms the basis for the special device known as the analog computer. This special computer allows the simulation of a wide variety of physical systems that are described by differential equations. Although there are many types of analog models (Figure 3–15), the analog computer is frequently used because of its convenience and versatility.

3.3 The role of engineering models

The primary role of the engineering model is served when it is used during the feasibility stage, the preliminary stage, the detail stage, or the revision stage of design to evaluate an idea that has been suggested to meet a specified need. Even if this were the only use for engineering models they would be one of the most valuable tools that a designer could use. Yet, engineering models play significant secondary roles in the industrial environment.

An engineering model enhances communication among engineers. For example, a computer program can be used by engineers to explain to each other the complex interactions present in an engineering system (Figure 3–16). If one picture is worth a thousand words, then one well-constructed model is worth several thousand words.

Figure 3–14 Analog computers are extremely versatile modeling tools in engineering design. (*Courtesy* EAI.)

A model is not only useful for communication among engineers within a design stage; it can also be used to transmit an idea from one stage of design to another. Frequently an engineering model will accompany the report output of one stage of design that is passed to the next stage. A physical or graphical model is nearly always used to communicate an idea from the design process to the manufacturing process.

An engineering model is often used to train personnel who will use, repair, or maintain a product once it is in service. A good example is the flight simulators that NASA has used to train its astronauts.

An engineering model is often used for archival documentation. Many companies maintain extensive libraries of the physical models and drawings used during the design and construction phases of a new project. These archives provide excellent reference information that often becomes extremely useful when questions arise during design revision.

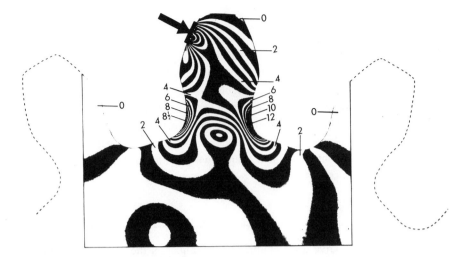

Figure 3–15 This photoelastic model of a gear tooth uses light interference patterns to identify lines of constant stress. (*Courtesy* The Measurements Group, Raleigh, NC.)

Figure 3–16 A computer is used for comparison of chemical reaction models. (*Courtesy* Tektronix, Inc.)

3.4 Summary

Engineering models are part of the equipment that designers use to perform their professional tasks. Often one idea will be studied by using many different types of models. The more models that an engineer understands, the better he or she will be as a designer. There is no substitute in the design process for a well-constructed model that is functional, economic, timesaving, and succinct.

Once an adequate model is available, the next step in the design process is evaluation. In the next chapter we will discuss some important aspects of the evaluation step.

PROBLEMS   **3.1**   Give three engineering examples of the following types of models:

(a) scale model

(b) diagram model

(c) mathematical model

(d) analog model

(e) pictorial model

3.2   Under what conditions would it be desirable to have a scale model that is larger than life size?

3.3   Name six general characteristics of all models.

3.4   Under what conditions would a photograph be considered an engineering model?

3.5   Can you find a mathematical model for predicting the following:

(a) The present value of a savings account containing "D" dollars that has been on deposit for "N" days at an annual interest rate of "p" percent with daily compounding.

(b) The current flowing in a resistor of "R" ohms that is connected to a battery of voltage "V".

(c) The total volume of a cone-shaped container with base of diameter "D" and height "H".

(d) The surface area of a spherical tank of radius "R".

3.6   Under what conditions would an analog model be more useful than a scale model of an actual system?

3.7 What is the difference between an analog computer model and a digital computer model?

3.8 Can you prepare a diagram model that depicts the sequence of events a student must go through at your university in order to sign up for a course?

3.9 Suppose you were asked to design a small lumber footbridge to cross a brook that is three meters wide. What type of model would you choose to use and why?

3.10 Suppose you were asked to design a new type of camping tent for a family of four persons. What type of model would you choose to use and why?

3.11 Suppose you wish to consider new arrangements of the furniture in your bedroom. What type of model would you use and why?

3.12 Under what conditions would it be justified for an engineering model to be more expensive than the object being designed?

REFERENCES

BEAKLEY, G. C. and H. W. LEACH, *Engineering: An Introduction to a Creative Profession*. 2nd ed. New York: The Macmillan Company, 1972.

BEAKLEY, G. C. and E. G. CHILTON, *Introduction to Engineering Design and Graphics*. New York: The Macmillan Company, 1973.

DOEBELIN, E. O., *System Dynamics Modeling and Response*. New York: Charles E. Merrill, 1972.

EIDE, A. R., R. D. JENISON, L. H. MASHAW, and L. L. NORTHRUP, *Engineering Fundamentals and Problem Solving*. New York: McGraw-Hill Book Co., Inc., 1979.

FLETCHER, L. S., and T. E. SHOUP, *Introduction to Engineering Including FORTRAN Programming*. Englewood Cliffs, N.J.: Prentice-Hall, Inc., 1978.

HILL, P. H., *The Science of Engineering Design*. New York: Holt, Rinehart & Winston, 1970.

KNOBLOCK, E. W., and J. N. ONG. *Introduction to Design: the Process of Problem Solving*. Milwaukee, Wis.: Spectra Ltd., 1977.

RUBENSTEIN, M. F., *Patterns of Problem Solving*. Englewood Cliffs, N.J.: Prentice-Hall, Inc., 1975.

The Exxon North Slope supertanker. (*Courtesy* Exxon USA.)

Design considerations

4

The designer must keep in mind that no matter how small or large, how simple or complex a design project may be, all aspects of the project must be investigated before the design is completed. Among the many factors which the designer must consider are the legal implications of the design, its economic impact, its social impact, and its effect on natural and energy resources. If the device is a marketable product, the designer also must be concerned with elements of manufacture or production. Whatever the design project, the engineer must ask: Will the design satisfy the appropriate standards and regulations imposed by law? Will the design adversely affect the environment? Will the design win public acceptance? Will the design meet the written and unwritten ethical standards of the engineering profession? Will the design be cost-effective?

Although the world appears to operate on a supply and demand basis, the development of a new device for the marketplace should be guided by specific standards and regulations, by professional and personal ethics, as well as by economic factors. Some techniques must be devised for evaluating the design in terms of its conformity to available standards and its potential long-range effects on the public and on the environment.

4.1 Standards and regulations

Numerous standards, laws, and regulations govern almost every phase of modern-day life. This is particularly true of engineering endeavors. Contrary to popular belief, these regulations are not purely a modern phenomenon. In the 18th century B.C., King Hammurabi of Babylon established regulations for the construction of buildings, stating that if a house collapsed and killed any of its occupants, the builder of the house would be put to death. Clearly, regulations imposed on persons in the performance of their work have roots deep in the past. Today, the standards and regulations are not as simple and straightforward. Their scope, however, extends to almost every phase of engineering work.

A *standard* is an exact value or concept that has been established by some authority or by agreement. The standard serves as a model or rule in the measurement of a quantity or in the establishment of some procedure. Standards establish applications and limitations for items, materials, processes, methods, and designs, as well as for engineering practice (Figure 4–1). A *regulation*, on the other hand, is a rule, ordinance, or law by which conduct or performance is controlled. Generally, a regulation is an accepted guide for uniform performance or a mechanism for controlling or governing a process.

Standards. The American National Standards Institute (ANSI) plays an important role in establishing standards of design and production in the United States.

Figure 4–1 Engineers consult standards for telecommunications equipment. (*Courtesy* Bell Laboratories.)

ANSI is a federation of over 900 companies and 200 trade, scientific, technical, professional, labor, and consumer organizations. American National Standards Committees within these organizations develop standards, and ANSI determines the establishment of consensus on the approval of these standards. ANSI publishes and disseminates standards for the manufacture of nuts, bolts, photographic film, spark plugs, traffic signs, pipe, wire, and just about every other commonly manufactured item. Currently there are approximately 8,500 American National Standards.

Although participation in ANSI and use of the standards it develops is voluntary, the adherence to these standards is beneficial to everyone. Consider the confusion if every manufacturer made light-bulb screw bases of different sizes or flashlight batteries of different diameters.

ANSI also arranges for effective representation of the United States in international standards organizations such as the International Organization for Standards (ISO). Much activity has taken place in recent years in developing international standards. This movement has resulted from international trade agreements and from the move to conversion, by the United States, England, Canada, and Australia, to the International System of Units (SI). At present, about 3000 ISO standards are in effect, and many new international standards are developed each year.

Many other organizations also maintain specialized codes or standards for various materials, products, and procedures. Some of the engineering professional societies have well-known standards. The American Society for Testing Materials (ASTM) has established procedures for testing all kinds of materials in various physical states. The American Society of Mechanical Engineers (ASME) is one of the major publishers of American National Standards, including the boiler and pressure vessel codes, nuclear power plant codes, and fastener and key codes. The Institute for Electrical and Electronic Engineers (IEEE) has developed specialized electrical codes for power plants. The American Society for Heating, Refrigeration and Air Conditioning Engineers (ASHRAE) has established standards for heating and cooling equipment. The American Society for Quality Control (ASQC) has established standards for quality control sampling procedures and tables. There are also a large number of trade associations which have developed standards or codes; among these is the American Iron and Steel Institute which has set standards for steel plate, I-beams, and channels.

Regulations and laws.   In the past decade, two agencies of the federal government have been established with responsibilities that strongly affect engineers and designers: the Environmental Protection Agency (EPA) formed by the National Environmental Policy Act of 1969, and the Occupational Safety and Health Administration (OHSA) formed by the Occupational Safety and Health Act of 1970. These agencies have been given the power to recommend and enforce standards and regulations related to the safety and health of workers (Figure 4–2), the quality of the environment in which they work, and the environment of the cities and communities throughout the country (Figure 4–3).

In addition to these agencies, the Consumer Product Safety Commission was established in 1973. It is the task of this Commission to determine whether a product poses an unreasonable risk of injury, to ascertain whether a mandatory safety standard would reduce that risk of injury, and to work with interested groups in the development of an appropriate standard. For example, the Commission recently determined that miniature Christmas-tree lights present unreasonable risks of injury from fire and shock, and that a consumer product safety standard was necessary to reduce or eliminate these risks. The Commission then initiated the development of a consumer product safety standard for miniature Christmas-tree lights and similar decorative lighting.

Product liability suits involving engineers and their companies have become increasingly common in recent years. At issue is whether the product meets the standards or regulations imposed by law. These cases may involve anything from poor quality control and use of inappropriate materials or components to improper design.

Figure 4–2 OSHA recommends and enforces safety standards for workers at plants such as this. (*Courtesy* Republic Steel.)

Engineers and their companies have the legal and moral responsibility to provide safe products. Even after a product has been sold, if a manufacturer finds that the product has dangerous defects in design or material or that the operating instructions are inadequate, improper, or ambiguous, the manufacturer must correct the defects or inadequacies. Over the past few years, design, material, and manufacturing defects have necessitated the recall of many consumer products— from automobiles to children's toys—at great expense to the manufacturers and accompanying inconvenience to the owners. A defect in design or manufacture may well provide sufficient grounds for a successful liability suit against the manufacturer, resulting in costly legal settlements and reduced sales.

The designer must stay abreast of all standards, laws, and regulations that pertain to the product he or she is developing. These rules provide direction for the designer and help to protect the manufacturer and the public at large by requiring that products be durable and reasonably safe for their intended use (Figure 4–4). Ignorance of the standards or laws does not absolve the designer for any

Figure 4–3 In the design and construction of cross country pipelines, the integrity of the environment must be maintained. (*Courtesy* Du Pont.)

Figure 4–4 Corrosion tests are conducted to assure that automobiles are durable under adverse conditions. (*Courtesy* Chrysler.)

errors or omissions in the resulting design. If it does not meet specified standards or regulations, the most creative design or the finest model may be utterly useless.

4.2 Patents

Most of the standards, laws, and regulations under which a designer works were devised in the recent past for the protection of the public. One basic law, however, has as its purpose the protection of the designer.

Article I, Section 8 of the Constitution of the United States states that "the Congress shall have the power . . . to promote the progress of science and useful arts, by securing for limited times to authors and inventors the exclusive right to their respective writings and discoveries." In patent law, the terms *discovery* and *invention* are used synonymously. These terms have become somewhat difficult to define precisely. They do, however, involve certain basic characteristics: some form of mental development, some experience or study and skill, and something which is new or a novelty.

The first Patent Act was approved on April 10, 1790, entitled "An Act to Promote the Progress of the Useful Arts." Thomas Jefferson, as Secretary of State, became the first administrator of the American Patents System. Because of Jefferson's limitless scientific curiosity and inventive mind, he served as the prime moving force in the promotion of the useful arts. Responsibility for granting patents was delegated to Jefferson as Secretary of State, to the Attorney General, and to the Secretary of War. Since that time, changes in the patent system have been brought about by a new Patent Act in 1836 and, more recently, the Act of 1953. All of these legislative changes have served to focus on the basic policies initiated by Jefferson.

Six major categories of patentable devices have been established by the Patent Acts: processes, machines, manufactured products, compositions of matter, designs, and hybrid plants. Each of these categories is defined in the Patent Acts and the patent notes which serve as interpretations of the law. It is not possible to patent natural products, ideas, data, natural laws or principles, or business techniques. The patent laws do not permit new patents for the substitution of materials, changes in form, proportion, or arrangement, or the exercise of ordinary engineering skill. Further, the substitution of one part for an equivalent part of a machine or process is not considered to be an invention.

Patent notes and federal statutes have established guidelines for the determination of inventions that are patentable and may be protected by a document which grants an exclusive right to their production, use, or sale for a certain time. There are many opportunities in engineering design to develop products or devices which are patentable (Figure 4–5).

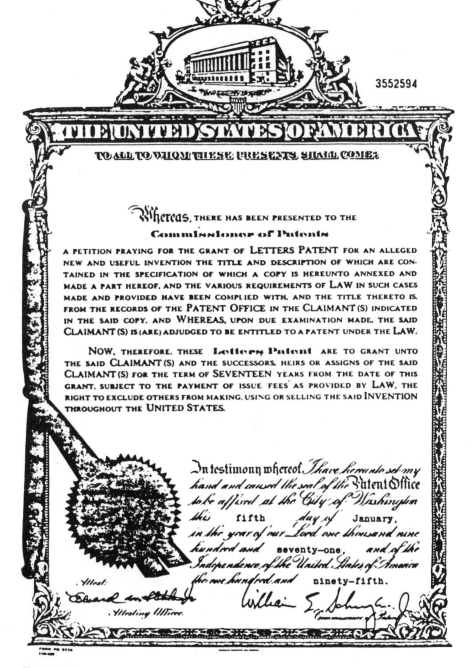

3552594

THE UNITED STATES OF AMERICA

TO ALL TO WHOM THESE PRESENTS SHALL COME:

Whereas, THERE HAS BEEN PRESENTED TO THE

Commissioner of Patents

A PETITION PRAYING FOR THE GRANT OF LETTERS PATENT FOR AN ALLEGED NEW AND USEFUL INVENTION THE TITLE AND DESCRIPTION OF WHICH ARE CONTAINED IN THE SPECIFICATION OF WHICH A COPY IS HEREUNTO ANNEXED AND MADE A PART HEREOF, AND THE VARIOUS REQUIREMENTS OF LAW IN SUCH CASES MADE AND PROVIDED HAVE BEEN COMPLIED WITH, AND THE TITLE THERETO IS, FROM THE RECORDS OF THE PATENT OFFICE IN THE CLAIMANT (S) INDICATED IN THE SAID COPY, AND WHEREAS, UPON DUE EXAMINATION MADE, THE SAID CLAIMANT (S) IS (ARE) ADJUDGED TO BE ENTITLED TO A PATENT UNDER THE LAW.

NOW, THEREFORE, THESE Letters Patent ARE TO GRANT UNTO THE SAID CLAIMANT (S) AND THE SUCCESSORS, HEIRS OR ASSIGNS OF THE SAID CLAIMANT (S) FOR THE TERM OF SEVENTEEN YEARS FROM THE DATE OF THIS GRANT, SUBJECT TO THE PAYMENT OF ISSUE FEES AS PROVIDED BY LAW, THE RIGHT TO EXCLUDE OTHERS FROM MAKING, USING OR SELLING THE SAID INVENTION THROUGHOUT THE UNITED STATES.

In testimony whereof, I have hereunto set my hand and caused the seal of the Patent Office to be affixed at the City of Washington this   fifth   day of   January, in the year of our Lord one thousand nine hundred and   seventy-one,   and of the Independence of the United States of America the one hundred and   ninety-fifth.

Attest:

Attesting Officer.

William E. Schuyler Jr.
Commissioner of Patents

Figure 4–5

84

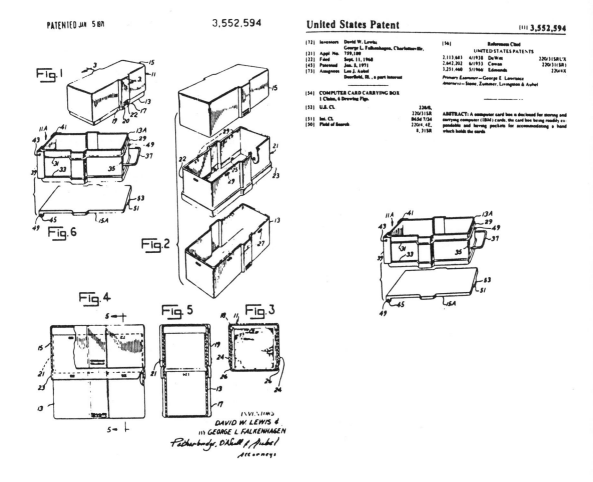

Figure 4–5 Patent assigned to D. W. Lewis, University of Virginia.

4.3 Economic considerations

Generally, an engineering design is developed for some purpose: to improve an old design, to establish a new product, or to develop a system which will improve the quality of life. The overall costs of the development of the product or process, as compared to the potential gain, will determine the economic feasibility of the product. The efficiency with which the technique, materials, and labor are combined to produce the product will help to determine the overall cost of the product.

Economic considerations are extremely important to the design process. Among the many diverse economic factors which must be considered in the design of any product or process are the marketplace (the supply and demand characteristics), the competition of other producers and suppliers, the maintainability and reliability of the product, the quality of the labor in the product, the fabrication techniques, the availability of materials and components, and the potential for obsolescence of the design, components, or fabrication facilities. The designer must stay abreast of new developments in his or her field in order to establish a basis for economic comparison.

The *marketplace* is an important part of the overall economic considerations. The design of a product must be geared to that segment of the population which needs or may want the product. The design must be sufficiently unique to have special attraction to this public. It is highly unlikely that a company would undertake the development of a product for which there is, or very likely will be, only a very limited market. The company which undertook the development of a new and modern buggy whip would likely fall on hard times.

The costs of the design, fabrication, and marketing must all be considered in developing the design of the product. Most engineering and manufacturing organizations have specialists familiar with the fabrication and marketing aspects of a product, but the designer must keep in mind these marketplace factors if he or she is to develop a successful design.

Competition is important to engineering design because it may provide the impetus for true innovation in the development of a product. Prudent designers will investigate the needs of the marketplace as well as the characteristics and capabilities of other products on the market before beginning a design project. Many large corporations, as well as the federal government, have design competitions for major products to assure that the ''best'' product is developed. In some federal programs, design competition provides different companies an opportunity to propose their design for a product. Such competitions are likely when the federal government wishes to purchase a unique vehicle such as the Space Shuttle or a new aircraft system. The production contract, then, is given to that company which provides the best overall program. This kind of competition sharpens the innovative and creative capacity of the engineer.

The *maintainability* and *reliability* of a product are of importance to the designer. Clearly it would not be feasible to develop an expensive product which cannot be repaired. The product should be designed for easy maintenance (usually with replaceable components) and consistent reliability. The designing engineer, then, must be familiar with the manner in which the product will be used, the types of replacement components readily available on the market, and the capability of the existing repair facilities and personnel (Figure 4–6). Effort devoted

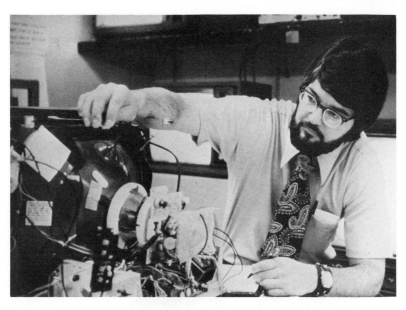

Figure 4–6 Television replacement parts and maintenance personnel are available throughout the country.
(*Courtesy* General Electric Company.)

by the designer to the improvement of the reliability and maintainability of a
product is effort well spent, as it will generally increase the market appeal of the
product.

The reliability of a product is directly related to the *quality of the labor* involved in its manufacture. A labor force which is not concerned about the manner
in which products are manufactured or the care with which components are assembled generally will produce an inferior final product. The designer should
keep in mind the capability of the labor force in the development of the design.

The designer must also consider the *fabrication techniques and facilities* to
be used in manufacturing a product (Figure 4–7). It is not difficult to design a
device that is aesthetically pleasing yet extremely difficult to fabricate. The design
should enable the simplest possible fabrication. Further, the longer the time required to fabricate a product, the higher the production costs and, consequently,
the lower the potential profit. It would hardly be feasible to design a new toaster
as a production item for your company when the company has been producing
high-pressure storage tanks and piping.

The *availability of material* also must be considered in designing a product.
The designer should take care that the materials specified are readily available on
the market and/or at the particular manufacturing location. The cost of materials

Figure 4–7 The fabrication of egg cartons requires sophisticated machinery. (*Courtesy* Diamond International.)

also must be considered. It is the goal of the designer to specify the best material with the least cost for use in the product. Low-quality material may meet the immediate needs of the manufacturer but may not last long. For example, a plastic gear in a drive train which must carry high torque loads probably would have very limited usefulness. Indeed, the substitution of lower-cost materials can sometimes cause trouble in the future and may actually cost more in the long run. The balance between quality and cost is a difficult one, but one to which the designer must be committed.

Possible *obsolescence* of fabrication facilities or components also influences the economics of the product. Certain production machines may become outdated and replacement parts unavailable. Competitors may develop production techniques that reduce manufacturing costs. Components available when the design was completed may become unavailable. New components or new production facilities may be developed after the basic design has been established. It is essential that designers and engineers keep track of the changing availability of components and facilities. For example, it would hardly be appropriate to design a camera for a specific film speed when new film developments result in higher and higher film speeds.

Economic considerations may become extremely complex, yet they are important factors in the development of a new design or process. Generally the engineer, the designer, or the company that provides a quality product for a reasonable sum is most likely to succeed in business.

4.4 Ethical and professional responsibilities

Ethics in the industrial world have become a topic of increasing concern. In the past, many people assumed that integrity and honesty underlay the products of most industries. As the number of product liability suits increases, it has become clear that the ethical and professional responsibilities of the engineer should be established. The engineer must develop designs and products which fulfill both legal and moral responsibilities. It is these responsibilities which make engineering a profession and the engineer a professional.

Ethics involve the behavioral concept of what is right and what is wrong. Without established standards of behavior, the world would be in chaos. Transactions of the business world must be completed in mutual trust and good faith.

There is a need for clearly defined rules of ethics whenever people interact. Certainly no profession can exist in isolation. Engineers as professionals must interact with others who rely on their services to fill their needs.

A number of well-developed ethical codes and canons have been devised by engineers and engineering societies in order to establish a degree of integrity and respect within the engineering profession. The ''Code of Ethics for Engineers'' adopted by the Engineers Council for Professional Development (ECPD) is perhaps the most generally accepted engineering code of ethics used today (Figure 4–8). Most of the codes and canons of engineering ethics deal with the relationship of the engineer to employer, to client, and to the public. In taking the initiative voluntarily to adhere to the ethical code of the engineering profession, the engineer adds respect, dignity, and prestige to the profession.

It should be noted that businesses and industries generally have published standards of conduct for their employees. These standards include stipulated violations, penalties, and appropriate enforcement agencies and processes. The engineer is responsible to his or her employer as well as to the engineering profession.

In addition to the professional code of ethics and possible employer standards, most states have engineering registration laws for the purpose of protecting the public from unqualified persons claiming to be engineers. Although laws differ from state to state, most states require the engineer to pass a written examination, show evidence of engineering competence, and agree to abide by state-approved codes of ethics. The examination portion of the registration usually con-

---

CODE OF ETHICS OF ENGINEERS

THE FUNDAMENTAL PRINCIPLES

Engineers uphold and advance the integrity, honor and dignity of the engineering profession by:

I. using their knowledge and skill for the enhancement of human welfare;

II. being honest and impartial, and serving with fidelity the public, their employers and clients;

III. striving to increase the competence and prestige of the engineering profession; and

IV. supporting the professional and technical societies of their disciplines.

THE FUNDAMENTAL CANONS

1. Engineers shall hold paramount the safety, health and welfare of the public in the performance of their professional duties.

2. Engineers shall perform services only in the areas of their competence.

3. Engineers shall issue public statements only in an objective and truthful manner.

4. Engineers shall act in professional matters for each employer or client as faithful agents or trustees, and shall avoid conflicts of interest.

5. Engineers shall build their professional reputation on the merit of their services and shall not compete unfairly with others.

6. Engineers shall act in such a manner as to uphold and enhance the honor, integrity and dignity of the profession.

7. Engineers shall continue their professional development throughout their careers and shall provide opportunities for the professional development of those engineers under their supervision.

Approved by the Board of Directors, October 5, 1977

Figure 4–8 Engineering code of ethics. (*Courtesy* Accreditation Board for Engineering and Technology [ABET], formerly Engineers Council for Professional Development [ECPD].)

sists of two parts, one dealing with engineering fundamentals that are common to all disciplinary branches of engineering, and one concerning professional practice. The engineering practice examination is oriented more to a specific discipline, with greater in-depth treatment of the subject.

Nearly all states have provisions for an "Engineer-in-Training" (EIT) status which allows a person to take a written examination on engineering fundamentals at the time of graduation from the undergraduate engineering program. Although the EIT offers no legal rights or privileges, this system of examination is a convenience to new graduates, since it allows them to take the fundamentals portion of their registration examination at a time when the material is fresh in their minds. Under this system, engineers may take another examination later in their career in order to complete the registration process.

Although the regulations differ from state to state, most states provide what is called "reciprocity." This means that an engineer who is registered in one state may become registered in another state provided the present registration is based on a procedure that is equivalent to or more rigorous than that of the new state.

Although it is possible to work for an engineering firm as a nonregistered engineer, a court of law generally will not recognize an individual as an engineer unless he or she is registered. In some states it is against the law to advertise oneself to be an engineer unless one is registered. In addition, there are certain engineering jobs involving public safety, public health, and public welfare that cannot be performed by unlicensed persons. It is likely that professional registration in the future will be even more important since it is often regarded as a measure of technical competence. For this reason every young engineer would be well advised to seek registration as quickly as possible after graduation.

4.5 Evaluation and assessment of design

Once the engineer has considered the standards, laws and regulations, the economic factors, and the ethical and professional responsibilities involved in the project, the design may be developed. Techniques discussed earlier, including the basic design method, idea generation, and model building, may be used in developing the design. Once the design has been developed, it is important that the engineer analyze the design to see that it satisfies the original objectives of the design program. It is difficult to evaluate philosophical objectives. It is necessary, therefore, to identify realistic objectives against which the results of the design may be measured.

In setting objectives, it should be realized that the assessment or evaluation of the performance of a design involves some uncertainty, particularly with new projects. For example, a design engineer for the VEX Company has just developed a square steering wheel for automobiles. This steering wheel appears to have all the assets of a round steering wheel, as well as several additional benefits. The design engineer must establish a means for assessing the performance of this new steering wheel. The function of a steering wheel (to determine the automobile's direction) is fairly clear; however, this function should be sufficiently flexible to allow for the evaluation of unforeseen problems—such as a corner of the steering wheel hitting the driver on the knee. The criteria for measurement of the objective should include accuracy and precision, as well as some form of performance measure. As the design is evaluated, problems may well arise. Problems such as fatigue failure at the corners of the steering wheel may be serious. It is necessary to ascertain what constitutes a major problem, how it should be solved, and what should be construed as an acceptable solution to the problem.

Several questions should be kept in mind as the engineer strives to design a product to meet certain objectives.

(i)   Objectives: Are the objectives established in terms of measurable quantities; if not, are the identified objectives sufficient?

(ii) Performance: Are there available performance measures, or can performance measures be identified that would be meaningful?

(iii) Testing: Is nondestructive testing possible, or is it necessary to damage the product in the testing process?

(iv) Interpretation: Can the data and information obtained be evaluated satisfactorily?

(v)   Modifications: Will modifications improve the performance of the design?

(vi) Long-Range Implications: Have any long-range problems or effects been identified?

The establishment of *measurable objectives* is perhaps the most important part of the design assessment. The engineer must consider whether the objectives are realistic, whether they represent the complete performance of the product, and whether they will be acceptable to the general public. An objective such as the design of a total energy residence system cannot be measured and therefore is of little practical value in the evaluation of a design. Once measurable objectives have been established, every designer should ascertain why a product tested does or does not satisfy those objectives. At this point, standards, laws, and regulations should be checked to assure that the design meets all requirements.

Performance is evaluated through *testing,* and data should be collected for all aspects of the performance or evaluation tests. The engineer should determine whatever may be learned from the various data measures (Figure 4–9), and how these data or resulting nondimensional parameters contribute to an understanding of the product performance. All results—even those caused by accident or erroneous measurements—should be accounted for. There are times when unexpected results do occur, and they may signal potentially serious problems with a product.

For example, in the design and development of a hand-held calculator, it is assumed that all mathematical functions should be performed precisely and without error. In checking all mathematical functions, it is noted that when multiplying 10 times 10, the newly developed calculator occasionally yields a solution of 101. All other functions are found to perform accurately. The designer must de-

Figure 4-9 Vibration and balance test instruments are used to analyze propeller and engine performance. (*Courtesy* Scientific-Atlanta.)

termine the reason for the erroneous calculation. The problem may result from a major defect in the electronic chip, or perhaps it may simply be from a faulty key contact.

All discrepancies in the data should be considered, and the *interpretation* of the data should account for these discrepancies. As a result of the performance test, the engineer should indicate what level of confidence may be placed on the result and on the performance of the product. If the pre-established objectives are not met, the engineer should make recommendations as to additional experiments, modifications in the design, or re-evaluation of measurable objectives so that the product and its goals can be made to coincide.

The recommendation of *modifications* is an important part of the design process. Rarely does a first design become a marketable product. Often the marketable product may be the 10th, the 15th, or even the 50th design of a product. Modifications are extremely important to improvements in both the performance and the marketability of a design (Figure 4–10).

The engineer also is responsible for ascertaining the *long-range implications* of the design, so far as possible. What will be its effects on the environment, working conditions, competition, the marketplace, and suppliers? Certainly, the engineer is not expected to foresee all potential long-range problems, yet any indication of possible difficulties should be explored. Unanswered questions should be brought to the attention of appropriate persons or organizations so that potential problems may be thoroughly evaluated and the product which eventually reaches the market is the best that it can be.

Figure 4–10 An engineer reviews the reproduceability of a modified printer component. (*Courtesy* General Electric Company.)

4.6 Technology assessment

The United States has prospered on a strong technological base, and new technological developments further strengthen that base. With each new development, however, there may be secondary and tertiary effects which pose unexpected and undesirable new problems for engineers and for society (Figure 4–11). There are many people who suggest that the new technological advances have not always been to the benefit of humanity.

For example, the automobile provides a tremendous opportunity for mobility and freedom. With the automobile, however, come such problems as a high dependence on liquid hydrocarbon fuels, a significant increase in air pollution, the necessity to establish large highway networks and systems, significant increases in fatalities due to automobile accidents, growth of suburban areas and deterioration of cities, and limited development of public transportation. Each of these secondary or tertiary effects has become a significant problem in the state and federal political arena. The recent energy dilemma further exacerbates the transportation problems in the United States.

The development of television has brought the world into the home and has served as an excellent educational media. With television, however, comes X-ray radiation, models of violence, and a decline in reading.

Asbestos fibers have been used in many beneficial applications and are an excellent fire retardant. These benefits, however, are now overshadowed by many possible health hazards.

Because secondary and tertiary effects of technological developments have

Figure 4–11 The secondary effects of major projects, such as the Smith Mountain Dam in Virginia, must be considered by design engineers. (*Courtesy* American Electric Power.)

become major problems, attempts to predict and evaluate such side effects have increased significantly. These evaluations are termed technology assessments, or technological forecasting and long-range planning.

The field of technology assessment is young, yet it has become increasingly popular both in engineering and in the political arena. In 1972, Congress established the Office of Technology Assessment to provide guidance to Congress in dealing with legislation involving technology. The mission of this office is to investigate the ramifications of technological developments in our country and to determine how these developments may affect the many different sectors of our society. An assessment is essentially an organized study of many effects which may occur in the development of a new technology or in the extension or modification of an existing technology. These studies serve as a base for the development of public policy options for policy-making bodies.

Technological forecasting and long-range planning deal primarily with the indirect or secondary effects of technology on our society and the environment. These evaluations provide some insight into possible long-range effects or the need for new technological developments. Technological forecasting and long-range planning are developed in two ways. One approach uses information about

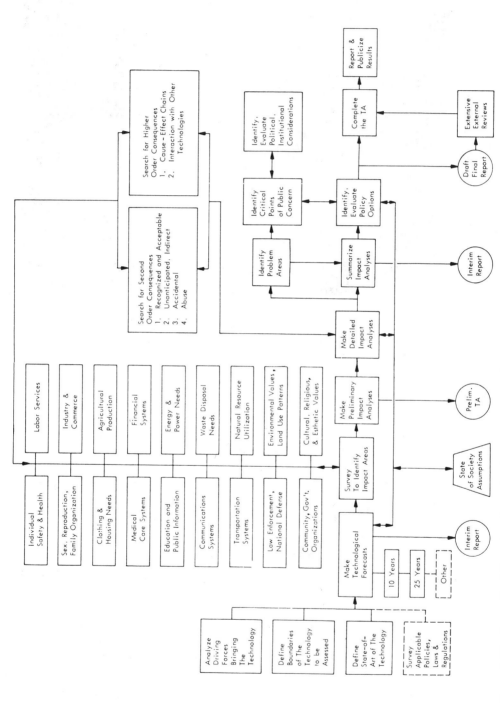

Figure 4-12 Generalized flow chart for technology assessment. (From *Technology and Social Shock* by Edward W. Lawless, © Copyright 1977 by Rutgers, the State University of New Jersey. Reprinted with Permission of Rutgers University Press.)

96

the past and present as a basis for focusing on the future. This approach is sometimes called exploratory forecasting. The second approach deals with the identification of some future goal or need, and works backward to current capabilities. This approach is sometimes termed the normative approach. Other forms of technological forecasting are essentially combinations or perturbations of these two major approaches. A generalized flow chart for technology assessment is provided in Figure 4–12. Note the many factors that are taken into consideration, as well as the possible time spans of the secondary and tertiary effects.

4.7 Summary

Many factors must be considered in developing a design for a new product, a new system, or a complex facility. The designer must be concerned with standards, laws, and regulations, with economic realities, and with ethical and professional responsibilities. The design process requires a great deal of thought, effort, and ability.

The field of engineering needs new designers with a wide variety of interests and training. Developing a new design can be an exciting endeavor and a rewarding part of a productive career in the broad field of engineering.

EXERCISES *Discussion topics*

4.1 Devise a standard for soft-drink containers, keeping in mind such considerations as the environment, safety, marketing, and production.

4.2 What sort of regulations or standards should be developed for elementary-school playground equipment?

4.3 Should there be standards or regulations for highway lighting and markers?

4.4 What evaluation and assessment factors should be considered in the design of an electric car?

4.5 What effect has the development of hand calculators and digital watches had on our daily lives?

4.6 Suppose that a new folding tennis racket has been developed, and your company has been asked to conduct an evaluation and assessment of the design. What factors would you consider?

4.7 With the advent of recombinant DNA comes an opportunity to improve the production of alcohol from biomass. What evaluation and assessment factors should be considered?

4.8 The rapid rise in microprocessors and minicomputers suggests the possibility of direct station-to-station hard-copy messages. What secondary and tertiary effects would such a message system have on our lives and on the postal service?

4.9 What do the following abbreviations stand for?

(a) ISO

(b) ANSI

(c) OSHA

(d) EPA

(e) ABET/ECPD

(f) OTA

4.10 What are some characteristics common to all professions?

4.11 What are the six categories of inventions established by the U.S. Patent Office? Give an example of each.

4.12 Give three examples of things which cannot be patented.

4.13 What is the difference between technology assessment and the design process described in Chapter 1?

4.14 What differences would you predict for the following products if they had been invented at a time when technology assessment was in use?

(a) the automobile

(b) highways

(c) the light bulb

(d) the airplane

(e) the bicycle.

4.15 What are the advantages to the engineer who uses technology assessment?

4.16 Why are standards and regulations important to the design engineer? to the consumer?

4.17 What agency of the federal government sets regulations with respect to the following design areas:

(a) noise levels

(b) air quality standards

(c) water quality standards

(d) food standards

(e) allowable radiation levels

REFERENCES

BEAKLEY, G. C., and H. W. LEACH, *Engineering: An Introduction to a Creative Profession*, 3rd ed. New York: The Macmillan Company, 1977.

DEGREEN, K. B., *Sociotechnical Systems: Factors in Analysis, Design, and Management*. Englewood Cliffs, N.J.: Prentice-Hall, Inc., 1973.

EIDE, A. R., R. D. JENISON, L. H. MASHOW, and L. L. NORTHRUP, *Engineering Fundamentals and Problem Solving*. New York: McGraw Hill Book Company, Inc., 1979.

FLETCHER, L. S., and T. E. SHOUP, *Introduction to Engineering Including FORTRAN Programming*. Englewood Cliffs, N.J.: Prentice-Hall, Inc., 1978.

GIBSON, J. E., *Introduction to Engineering Design*. New York: Holt, Rinehart and Winston, Inc., 1968.

KING, W. J., *The Unwritten Laws of Engineering*. New York: The American Society of Mechanical Engineers, 1944.

KRICK, E. V., *An Introduction to Engineering: Methods, Concepts, and Issues*. New York: John Wiley and Sons, Inc., 1976.

LAWLESS, E. W., *Technology and Social Shock*. New Brunswick, N.J.: Rutgers University Press, 1977.

TEICH, A. H., ed., *Technology and Man's Future*, 2nd ed. New York: St. Martin's Press, 1977.

TUSKA, C. D., *Patent Notes for Engineers*, 7th ed. New York: McGraw-Hill Book Company, Inc., 1956.

VAUGHN, R. C., *Legal Aspects of Engineering*, 3rd ed. Dubuque, Iowa: Kendall/Hunt Publishing Company, 1977.

WOODSON, T. T., *Introduction to Engineering Design*. New York: McGraw-Hill Book Company, Inc., 1966.

High Voltage Transmission System. (*Courtesy* General Electric Company.)

Engineering communications

5

Communication is essential to life as we know it. It is a means for sharing information and may be classified as a basic human activity. The history of communication is the story of a shrinking world. As far back as 3000 B.C., a picture language called hieroglyphics was developed by the Egyptians. This form of communication involved sketches or symbols inscribed in clay or rock to convey a message. In 1500 B.C., an alphabet was devised by the Semites, and details of activities were recorded. By 600 B.C. the Babylonians had established libraries dealing with subjects associated with their culture. In 105 A.D. the Chinese used paper and ink to record the events of the day and later developed a block-printing technique for reproducing written material. In the 1400's Gutenberg invented movable metal type, and the first printing press in the western hemisphere began operation in Mexico City in 1539.

The invention of telecommunication systems such as telegraph, radio, telephone, and television have revolutionized communications, brought the world closer together, and accelerated development (Figure 5–1).

Engineering is a profession which draws men and women from diverse backgrounds to play significant roles in the development of technology. Engineers generally work together; few engineering accomplishments have been achieved by individual engineers, nor exclusively by engineers of a single discipline. Communication is extremely important to engineering, whether for reporting new in-

Figure 5–1   Twin 10 m earth stations relay voice and data service via satellite. (*Courtesy* Scientific-Atlanta.)

ventions or designs, or for working with others on applications of engineering fundamentals to specific problems. Indeed, communication plays a significant role in all phases of engineering from the stating of a problem to the reporting of a solution.

5.1   Importance of communication

The purpose of communication is to convey an idea, problem, or solution to others. The knowledge base of today's engineers is built on the accomplishments of other engineers and scientists. The student engineer, then, must rely significantly on communications of others in order to acquire the training necessary for entry into the profession. Similarly, practicing engineers will continue to add to the knowledge base by communicating their work and accomplishments. Careful development of communication patterns and skills is extremely important to success in engineering.

In engineering design, communication plays a unique role because of the need to convey creative ideas and designs to others for implementation. Generally, communication in engineering design involves a combination of oral, written, and graphic instructions. Clarity and accuracy are the two most important elements in any engineering communication; without them, the communication may well be worthless. In the design of a new product requiring machine work, for example, precise dimensions of the components must be clearly stated, as must the material to be used, the finish expected, and the tolerance necessary for matching other parts.

The designer must develop the ability to communicate clearly, accurately, and concisely all the information to be conveyed. The content and form of communication—oral, written, graphical, or some combination of these—will depend upon the type of information to be conveyed and the audience to whom it is directed. Certainly a new design would be described in different terms when seeking approval of the board of directors of a company than when instructing a machine shop as to its manufacture. To be effective, then, the engineer must be able to communicate ideas and designs through drawings, spoken words, and written words to a wide variety of audiences.

5.2   Graphical communications

Among the first forms of graphical communication and earliest known forms of written communication were the hieroglyphics, or pictographic writings, of the ancient Egyptians. These pictographic writings were conventionalized pictures used to record stories and events. Although hieroglyphics formed the basis of the Egyptians' alphabet, the pictograph could convey a great deal more meaning. Even today, pictorial or graphical communications are considered by many the most efficient means of conveying concrete information.

The use of sketches, drawings, diagrams, or charts is an integral part of almost every aspect of an engineering project (Figure 5–2). In engineering design, pictographs or drawings can convey a great deal more information, in more vivid and realistic detail, than can be transmitted efficiently in any other way. For this reason, engineering drawings form the focus of most design projects.

Sketching. Sketching is used every day to convey many kinds of information. The police sketch the relative location of the automobiles involved in an accident. The interior decorator sketches furniture arrangement in a room or a house. The architect sketches various house designs. Sketching is equally important in engineering.

If a design idea is to be useful, the engineer must be able to convey its essence to others. After an idea is hatched, it is important to get the idea on paper, generally as a rough freehand sketch. Such a sketch provides an approximation of a device, object, or part, with emphasis on shape or concept rather than exact size (Figure 5–3). The first freehand sketch provides the engineer an opportunity to look over the idea, make modifications or rearrange components. Several additional freehand sketches may be made before the idea is finalized. The freehand sketch with a reasonable amount of detail serves as the means of communicating

Figure 5–2 The development of an engineering drawing. (*Courtesy* American Electric Power.)

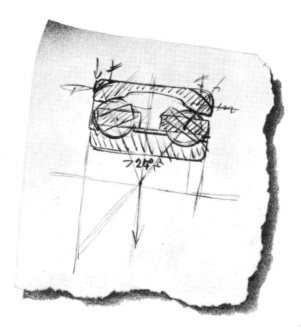

Figure 5–3 A typical freehand sketch of a ball bearing.
(*Courtesy* SKF.)

an idea to a supervisor, colleague, draftsperson, technician, or machinist. The preparation of a clear, accurate sketch, then, is extremely useful in communicating the basic characteristics of an idea to others. Techniques and guidelines for sketching will be discussed in more detail in Chapter 7.

Engineering drawings. Although sketches may be useful for conveying ideas, they generally are not sufficient for the actual fabrication of a device or part. Normally, these sketches are used by the draftsperson or technician to develop a set of working drawings or engineering drawings which may be used in the manufacture of the device. Numerous details must be made available to the machinist who will fabricate the device. Of particular importance are such details as dimensions (with tolerances), surface finish, type of material, number of parts, type of threads, and class of fit. A number of drawings may be necessary to represent the components of a complete design. It may be necessary to prepare individual drawings of each component and subassembly (Figure 5–4). Often, complete assembly drawings are provided for use in visualizing the complete device (Figure 5–5).

Engineering drawings normally are prepared with appropriate instruments

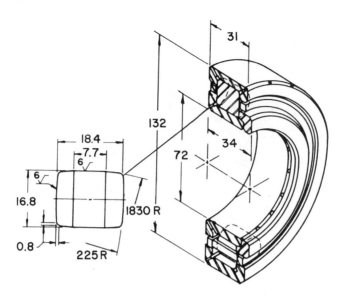

Figure 5–4 A typical subassembly drawing of a roller bearing.

(including computers) to provide clear, sharp drawings which may be reproduced for distribution to the machinist, as well as to consumers who might request details of the device. The techniques used to develop these drawings—termed *graphics, engineering drawing,* or *mechanical drawing*—are extremely important to engineering design and will be covered in detail in Chapter 7. The accurate representation of ideas through detailed drawings is essential to the development of a successful design project.

In addition to sketches and detailed drawings, the engineer may use other forms of graphical communications, such as schematic diagrams and charts, to communicate his or her ideas.

Schematic diagram. Diagrams are used for many purposes. They may describe movements in football or basketball games, geometrical figures, or the outline of some component. The schematic diagram frequently is used in engineering to aid in the communication of ideas. In such a diagram, an engineering process is represented pictorially through a series of abstract figures or pictures. (The propulsion system of a ship is shown schematically in Figure 5–6.) In schematics, or pictorial representations, the size and shape of a component are not retained, yet the essential logic of the process or idea is accurately conveyed.

Schematic diagrams may be simplified through use of conventional symbols

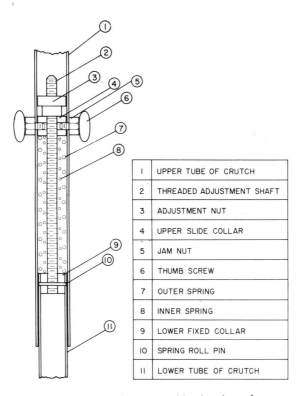

I	UPPER TUBE OF CRUTCH
2	THREADED ADJUSTMENT SHAFT
3	ADJUSTMENT NUT
4	UPPER SLIDE COLLAR
5	JAM NUT
6	THUMB SCREW
7	OUTER SPRING
8	INNER SPRING
9	LOWER FIXED COLLAR
10	SPRING ROLL PIN
11	LOWER TUBE OF CRUTCH

Figure 5–5 A complete assembly drawing of a pogo crutch.

for standard components such as resistors, transistors, pumps, tanks, turbines, or motors. Generally, these schematic diagrams portray the process but do not provide actual size or location of components (Figure 5–7). It is much easier to visualize the process through a schematic rather than in a photograph of the actual facility. Diagrams of this type are used throughout engineering under such diverse names as electrical network or circuit diagrams, flow diagrams, or building air handling and plumbing diagrams.

Charts. In addition to sketches, detailed drawings, and schematics, the engineer may use other graphical techniques to relate ideas. In engineering analysis, a mathematical determination of the statistical characteristics of a set of data does not always provide sufficient insight. For this reason, there may be a need to present numerical data in other forms for easier interpretation, particularly for non-technical audiences. Graphical presentation of data through the use of various

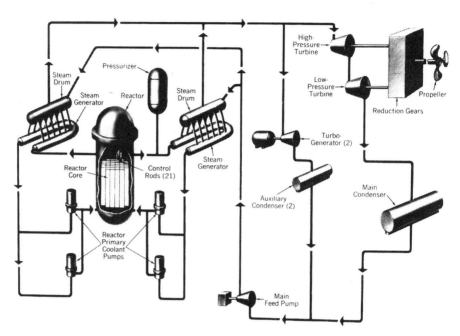

Figure 5–6 A schematic diagram of the propulsion system of the nuclear ship Savannah. (*Courtesy* Babcock and Wilcox.)

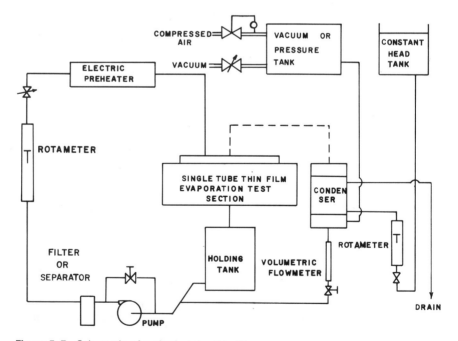

Figure 5–7 Schematic of a single tube thin film evaporation test facility.

graphs or charts simplifies the determination of transient characteristics of a data set and helps to make the information clearer to persons not trained in engineering.

The engineer may use a variety of pictorial charts to convey information. A line graph is used to show a trend or range of variation in a parameter (Figure 5–8). Bar graphs may take many forms and are often used to show changes in a parameter from year to year (Figure 5–9). Circular charts, sometimes called pie charts, show how a quantity is divided into various groupings. The distribution of energy use in a community is shown by means of a pie chart in Figure 5–10. Nomographs are extremely useful for portraying the relationship among several different variables (Figure 5–11). Other types of charts include navigational charts used by pilots, weather charts or maps, segment charts, and organizational charts (Figure 5–12).

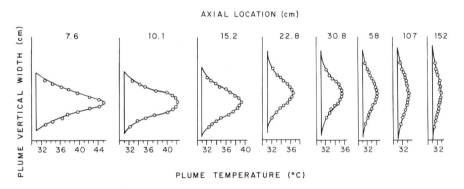

Figure 5–8 Thermal profiles of a smoke plume downstream from a smokestack.

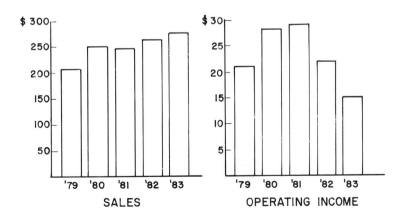

Figure 5–9 Typical income bar graph for the ZIPP Company.

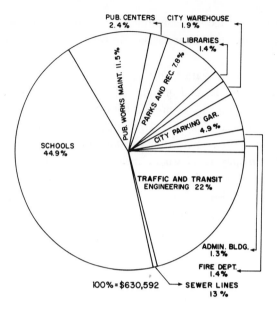

Figure 5-10　Pie chart showing distribution of energy cost in public buildings of a small community.

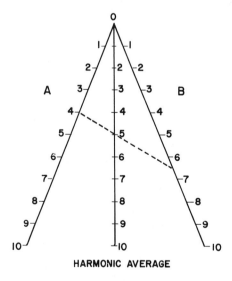

Figure 5-11　A simple nomograph for finding the harmonic average.

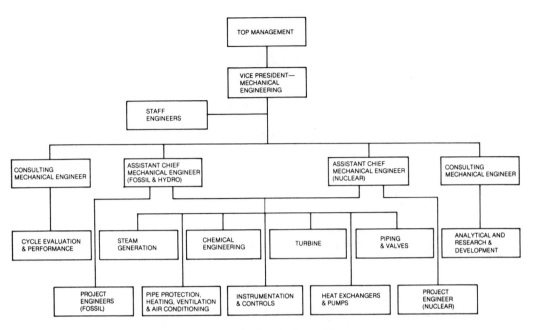

Figure 5–12 A typical organizational chart. (*Courtesy* American Electric Power.)

Of all forms of graphical communication, engineering drawing is the most important and most useful for engineers. Because a thorough mastery of the techniques of engineering drawing is essential to engineering design, this topic will be treated in detail in a later chapter.

5.3 Written communications

An engineering project is not complete until it has been reported. To provide a clear, thorough description of a process or project, the engineer combines writing skills with mathematics, drawings, and graphs. An engineering design project involves effort, time, and money. The final report on the project should be clear, concise, and thorough. Such communications should be as accurate and error-free as the technical aspects of the project are. It is essential that every engineer strive for quality communication, just as he or she strives for quality in technical work (Figure 5–13). The success of a project depends in part upon the technical capabilities of those involved in the investigation. It also depends upon the organizational and communication skills of the members of the engineering team.

Figure 5–13 Preparation of a design report. (*Courtesy* General Electric Company.)

General guidelines for written communications. Although this text is not intended to serve as a resource book on creative or reportive writing, it still may be useful to consider a few guidelines for content and style in written communications.

1. Identify a subject which is narrow, precise, and well-defined. The broader the subject of the written communication, the more difficulty the writer will have in conveying concrete and significant information. A well-defined subject facilitates the preparation of a meaningful document.

2. State the scope and limits of the subject under consideration. The writer should establish clearly what is and what is not to be included in the communication. Any conditions or limitations should be noted.

3. Keep the objective of the communication in mind at all times. Written communications are prepared for many different reasons. Perhaps the writer seeks approval of a supervisor or board of directors in order to pursue a particular project. Perhaps a new design is to be described for a patent application. The aim of the communication will dictate the style and content of the document.

4. Keep in mind the audience to whom the communication is directed. A communication prepared for a nontechnical audience will differ from that pre-

sented to a technical audience. It may be necessary to provide background material or to choose a vocabulary suitable for the audience. Jargon should be avoided, and content and style should be tailored to meet the needs of the audience.

5. Write clearly, concisely, and accurately. A communication is intended to convey information. The measure of a successful communication is the clarity, brevity, and accuracy with which it achieves this purpose.

6. Avoid generalities and biased interpretations. Engineers rely heavily on evidence derived from investigations. Caution should be used in drawing conclusions from these investigations. Statements should be qualified as to range of application and degree of uncertainty. Interpretation of facts should be made impartially and not directed toward a desired result. The ethics and professionalism of the writer will determine the accuracy with which information is interpreted and presented.

7. Cite appropriate sources for information used. It is important that facts be referenced as to source and that credit be given where credit is due. References provide the reader a perspective on the information base as well as a means for pursuing further information on the data, design criteria, or opinions presented.

8. Check for errors of commission or omission. The writer should review the written material to ensure its accuracy in form and content. The document should be checked to be sure that no important information has been left out, and that the information included is both necessary and reliable.

Types of written communications.   Some of the many types of engineering communications are noted in Table 5–1. These forms of communication are a part of everyday engineering activity and will become familiar as you embark on an engineering career.

A *memorandum* speaks to a specific subject and provides a quick communication to others concerned with the problem. Memorandums usually consist of several paragraphs and may be in the form of a *letter report*.

The purpose of an engineering *proposal* is to describe to others what you or your organization would like to do and how you would like to do it. Proposals may be directed to others within your organization or to an outside funding agency.

Technical reports generally are published by industrial corporations, universities, or federal government laboratories and facilities. These reports describe a process, review a specific subject or problem area, or give general information

Table 5–1   Types of Written Communications

Memorandum	Technical Note
Letter Report	Textbook
Proposal	Reference Book
Technical Report	Book Review
Bibliographic Review	Article Review
Journal Article	

for others working in the same area. Internal technical reports are sometimes considered to contain proprietary information and are not available for publication or distribution outside the organization which publishes them.

In order to provide timely information on a specific subject, a *bibliographic review* or survey might be prepared. Such reviews may be authored by an individual interested in the area, by a company or governmental agency, or by a library research organization. Reviews are sometimes published in subject area indexes or journals. Bibliographic listings of recently published work are readily available in most libraries and may be obtained by computer searches of current literature.

When an investigation has been conducted or a new design developed, it may be reported as an *article* in one of the technical or trade journals. Publication in journals provides a wide distribution of the work, since many individuals, companies, and libraries subscribe to such periodicals. *Technical notes,* also published in journals, generally are shorter and deal with a small portion of a problem area.

In order to provide a complete overview of a subject, a *textbook* or professional *reference book* may be written. These books provide basic background information, analysis techniques, and interpretations of existing problem areas. Engineering handbooks provide a great deal of information and are used frequently by the practicing engineer.

To assist the engineer in dealing with the plethora of material published each year, brief *reviews* of books and articles are published regularly in journals and trade magazines.

In engineering design, proposals and design reports are particularly important and will be discussed more thoroughly in the following sections.

5.4   Engineering proposals

Before beginning a design project, the prudent engineer develops a proposal for the project to be undertaken or the area to be explored. A proposal is a plan or scheme describing how a problem might be approached, or how a subject might be investigated. The proposal is often the basis for acceptance or rejection of a project by a supervisor, a board of directors, or an institutional or federal funding source. An engineering proposal usually explains the objective of a particular study, the equipment and workforce required, the anticipated results, and the length of time needed for the investigation. Frequently, a cost estimate is also necessary, particularly if a significant amount of support is required.

The engineer should consider all aspects of the planned project before preparing a written proposal. What previous work has been done in the area? What are the objectives of the present study? What manpower is required? Will the project be worth the funds invested in it? The engineer should anticipate the needs and concerns of the supervisor, industry, or organization to whom the proposal is being submitted.

Table 5–2 identifies the various components of a typical project proposal. The *title page* should include the concise title of the project, the names of the investigator(s) or project director, the person or organization to whom the proposal is submitted, and the date submitted. Often, the time period proposed for the project and the costs involved also are presented on the title page.

Following the title page should be a brief summary of the objectives of the project and the expected results. This may be termed the *abstract,* the summary, or the executive summary. It includes in clear, concise form all the salient points of the proposal.

Table 5–2 Components of a Typical Project Proposal

Title Page	Experimental or Design Procedure
Summary or Abstract	Personnel
Background and Literature Review	Time Schedule
Proposed Investigation Objective	Cost Estimate
Required Equipment and Instrumentation	Bibliography

The next section of a typical engineering proposal presents *background* material and a review of the relevant literature (Figure 5–14). This is an extremely important section of the proposal, as it provides the justification for the proposed investigation or design project. A brief history of the subject should be presented, and the relevant theoretical, experimental, or design work of other investigators should be discussed briefly. Areas which appear to warrant further investigation, and in which it is possible for the proposer to make a contribution, should be identified.

Following the background or literature review section, the *proposed investigation* should be described in detail, indicating the scope of the subject to be studied, the specific problem areas to be resolved or investigated, and the manner

Figure 5-14 A literature search to develop information for the proposal. (*Courtesy* Luther Gore, University of Virginia.)

in which the tasks are to be accomplished. This section should include a brief discussion of any theoretical, experimental, or design work planned, and how this work relates to the objective of the proposed investigation. It is essential that the engineer provide a clear, concise statement of what is to be accomplished, how it is to be accomplished, what is needed to accomplish it, and the value of the results anticipated. The practical value and anticipated benefits of the project also must be explained with sufficient justification to warrant support.

The next section of the proposal should list the *facilities required* (Figure 5-15) as well as the equipment, instrumentation, and design models needed for the investigation. This section might also include space requirements, particular components which may be manufactured, purchased, designed, or developed, and any necessary changes in existing facilities. It is not always possible to anticipate all the needs of the proposed project. The engineer should, however, consider every aspect of the experimental program and be as thorough and realistic as possible in the estimates.

The *method of investigation,* including the experimental or design procedures, should be presented in the next section. The engineer should present an outline of the total design program, indicating the type of information to be obtained, the procedures for analysis and design, the specific approaches to be used,

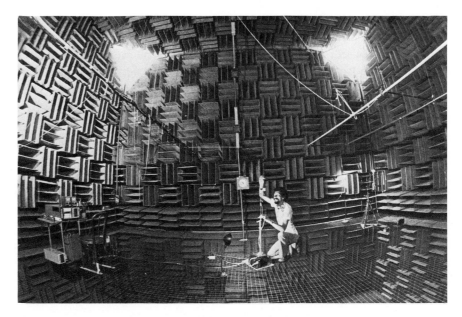

Figure 5–15 Sophisticated facilities like this anechoic chamber are necessary for development of a directional microphone. (*Courtesy* Bell Laboratories.)

an idea of the expected results, and the range of test parameters or design considerations that will be used.

A proposal is not complete without a description of the *personnel* to be involved in the project. The principal investigator should be identified. The qualifications of each major participant in the project should be provided, including experience, special expertise, prior involvement in similar programs, and the amount of time each plans to devote to the project. Other manpower requirements also should be identified.

It is important that a *time schedule* be prepared for the completion of the proposed project. This may be done by determining the time to be allotted for each portion of the investigation from the proposal to the final report, by PERT–CPM or GANT Charts, or by other methods of time scheduling. Techniques for developing such time lines will be discussed in Chapter 9. It is not always possible to follow a time schedule precisely. Problems may arise which necessitate reconsideration of time allotted for a particular phase. It is important, however, to establish a time frame for completion of the project.

The *cost estimate* is of substantial importance, particularly to a funding agency. This section of the proposal should list the cost of the material, instrumentation, or facilities which must be purchased as well as the cost of any necessary modifications to existing facilities. The number of hours required for machine-shop service, fabrication service, design facilities, technicians or laboratory-aid support, computer time, drafting, and so forth, should also be estimated. It is useful to indicate, too, the amount of professional time to be spent on the project. The total cost of the project may well be the determining factor in acceptance or rejection of the proposal. The funding source will assess whether the cost of the project is commensurate with the value of the results anticipated. At this point in the preparation of a proposal it is essential to consider again the audience to whom the proposal is directed and the value, in that audience's eyes, of the work to be conducted.

The *bibliography* is usually the final section of the proposal. In the literature review section, other theoretical, experimental, or design studies on the same or similar topics were discussed. It is necessary and appropriate to include in proper bibliographic style, those references or sources that are pertinent to the proposed design or investigation. They may be books, journal articles, manufacturers' catalogs, colleagues, or technical reports and documents. An example of typical bibliographic entries is provided in Table 5–3.

Table 5–3 Typical Bibliographic Entries

ACKLEY, ROBERT A. *Physical Measurements and the International (SI) System of Units,* Third Edition. San Diego: Technical Publications, 1970.

"AISI Metric Practice Guide: SI Units and Conversion Factors for the Steel Industry." Washington, D.C.: American Iron and Steel Institute, 1975.

BURTON, WILLIAM K., Ed. "Measuring Systems and Standards Organizations." New York: American National Standards Institute, Inc., 1970.

CHISWELL, B., and E. C. M. GRIGGS. *SI Units.* Sydney: John Wiley & Sons, Inc., 1971.

"Conversion to the Metric System of Weights and Measures," Hearings before the Subcommittee on Science, Research, and Technology of the Committee on Science and Technology, U.S. House of Representatives. Washington: U.S. Governmental Printing Office, 1975.

DE SIMONE, DANIEL V., Ed. *A Metric America: A Decision Whose Time Has Come.* National Bureau of Standards Special Publication 345, July 1971.

MECHTLY, E. A. "The International System of Units: Physical Constants and Conversion Factors," Second Revision. NASA SP-7012, 1973.

PAGE, CHESTER H., and PAUL VIGOUREX, Eds. "The International System of Units (SI)," National Bureau of Standards Special Publication 330, July 1974.

"SI Units and Recommendations for the Use of Their Multiples and of Certain Other Units," International Organization for Standardization, ISO 100, 1973.

"Standard for Metric Practice," E 380-79, American Society for Testing and Materials, 1980.

STIMSON, H. F. "The International Temperature Scale of 1948," *NBS J. Research,* Vol. 42, March 1949.

The thorough preparation of a proposal will provide additional insight into the problem and will improve the chances of the proposal's being approved. The proposal itself tells a great deal about the individual preparing it and the quality of the work that will be conducted. Accuracy, clarity, and a professional approach will help ensure success.

5.5 Engineering reports

Once a proposed project has been approved and the analytical, experimental, or design work has been completed, it is essential that a report be prepared to describe the results of the investigation. Careful thought must be given to the logical organization of the material to be presented in the final report. Consideration of the subject and anticipation of the readers' needs and questions usually will indicate to the writer the type of organization required. Although the form of an engineering report may vary, certain essential elements are presented in Table 5–4.

The first step in the preparation of a report is the development of a preliminary outline. An orderly plan for the report must be established, keeping in mind the overall objective of the project and of the report. A thorough, comprehensive outline will be of immeasurable help in the preparation of the report.

The report begins with a *title page*. The title of the technical report should tell accurately and clearly what the report is about and should be as specific as possible, within reasonable limitations of length. The title page should include the name(s) of the author(s), and the date that the report was prepared. It might also be appropriate to indicate the duration of the project, the laboratory or organization where the work was completed, and the author's business address or affiliation.

An *abstract* follows the title page, stating clearly and briefly the purpose of the project. The experimental plan used and the principal findings should be accurately but succinctly summarized. The major point of the program and appropriate conclusions should also be included. Although read first, the abstract should be written last to be sure that it accurately reflects the content of the report. Generally an abstract should be no more than one page, or approximately 200

Table 5–4 Components of a Typical Engineering Report

Title Page	Results
Abstract	Discussion
Introduction	Summary
Facilities or Techniques	References or Bibliography
Analytical, Experimental, or Design Procedure	

words. Since it is often extremely difficult to condense the goals, methods, and conclusions of a project into a single page, all the literary skills of the writer must be mobilized for this effort.

The next section of the report is the *introduction*. A good introduction states the problem clearly. It presents the background and appropriate justifications of the work, and describes the approach used to complete the project. It should indicate the significance of the work and outline what has been accomplished.

The *techniques* or *facilities* used in the investigation are described in the next section. Details of the facilities and instrumentation—including size, performance, range, and other test parameters—should be described. Specific techniques or approaches used in the design process should also be discussed. Frequently it is useful to provide sketches or photographs of the equipment to familiarize the reader with the specific facility used. A schematic or photograph usually will convey information more clearly than words alone.

The next section of the report describes the *analytical, experimental,* or *design procedure* that has been used. In this section the project plan is described, the experimental procedure is detailed, and/or the basic analytical approach is reported. Specific steps in the design process should also be described. Enough detail should be given so that other experienced workers familiar with the field could repeat the work. For experimental work, the method of preparation of models should be described, and the procedure used for testing should be reported (Figure 5–16). Specific details concerning the accuracy of primary measurements, the types of measurements made, and the overall reliability of the investigation should be included. For an analytical investigation, this section of the report might detail the techniques used, the computer time required, the stability of the computer program, or the number of nodes used in a finite element analysis. In a design project, the method of design should be discussed, including the various steps taken to arrive at the final design. Design considerations, cost factors, material selection, production techniques, and general layout problems should be described. All of these details are important to the reader seeking to understand the project.

The *results and discussion* section is the heart of the report. Ordinarily the presentation of results and the discussion of their significance are combined into a single segment of the report. In some situations, however, it may be appropriate to separate them into two distinct sections or to use a chronological approach. Whatever approach is selected, the following points should be carefully noted:

(i) Only relevant information should be included. Equations, figures, sketches, drawings, and tables should be introduced when necessary for clarity. Listings and summaries of data should be presented in appropriate appendices.

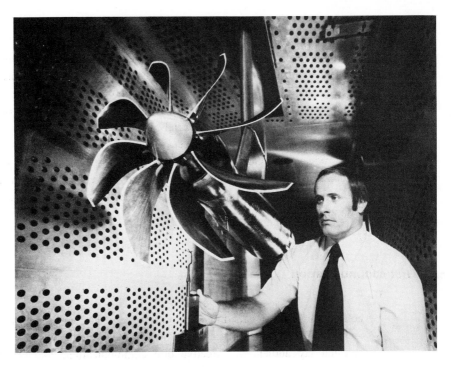

Figure 5–16 A model of a new aircraft propeller ready for testing in a wind tunnel. (*Courtesy* NASA.)

(ii) All numerical data should be reported in accepted systems of units, and appropriate notations and abbreviations should be used.

(iii) The major features and limitations of the investigation should be pointed out, and the results should be interpreted and compared with existing information. The engineer must evaluate and present the results in an objective, ethical, and professional manner and must not draw conclusions that cannot be justified by the design, data, or results obtained. Where uncertainty exists, firm conclusions should not be drawn, but rather the range or limits within which the conclusions may be valid should be stated.

The concluding section of the main body of the report, the *summary*, reviews the objectives and progress of the project and interprets the major conclusions. The summary provides an opportunity to point out secondary effects which would not be presented in the abstract. If the problem under study has not been completely resolved, further study may be suggested. Potentially productive areas of future research may be indicated. The clarity and precision of the summary will directly reflect the professional quality of the project.

In all reports, it is essential to include a citation for any references or library materials which have been consulted or used. *References* usually are cited in the order in which they are referred to in the report. A standard form of reference citation is shown in Table 5–3. Sufficient information must be given to enable easy retrieval of the reference material by the reader.

The reputation of the engineer is based in large part on the quality of the work produced. Written communications in the form of proposals, reports, memorandums, technical notes, and journal articles are very often the measure of that quality. A successful engineer must carefully cultivate the art of written communication.

5.6 Oral communication

The engineer is often called upon to present his or her work orally to co-workers, supervisors, professional society meetings or conferences, boards of directors, or interested lay audiences. Effective oral communication, like effective written communication, requires careful planning. Because of necessary time limitations, it is absolutely essential that the presentation be carefully planned and rehearsed. The use of video equipment to rehearse and critique presentations can be very helpful. A general outline for oral presentations is shown in Table 5–5.

In oral communication, it is important to integrate the audience, the purpose of the presentation, and the subject matter to be discussed. The language should be matched to the audience (Figure 5–17). Certainly the content and presentation of a talk on microprocessors would be far different for a group of electrical engineers than for a local community forum. Whatever the audience, the use of equations and highly technical terms should be limited to the ones absolutely nec-

Table 5–5 Components of a Typical Oral Presentation

Statement of the Objective
Summary of Conclusions and Recommendations
Development of the Subject
Presentation of Important Results
Conclusions and Recommendations
Wrap-Up
Acknowledgments
Questions

Figure 5–17 The style and content of an oral presentation will depend upon the audience to whom it is directed. (*Courtesy* Republic Steel Corporation and American Electric Power.)

essary to make a point or to establish a basis for discussion. Appropriate visual aids and supporting exhibits should be used whenever possible for clarity and emphasis.

The first step in an oral presentation is to provide a succinct *statement of the objective* of the project and/or presentation. The primary purpose and any secondary purposes of the project should be identified, and the scope of the subject area should be indicated. It is necessary to be concise and to touch only on the major points so as not to confuse the audience. The introduction to the talk is extremely important because it stimulates the audience's interest in the work that has been conducted.

The second portion of the talk should be a brief *summary of the conclusions and recommendations* of the study or investigation. These should be presented briefly and clearly. The engineer may find it useful to enumerate his or her points by number—that is, first . . . , second . . . , third. . . . These are the major points that you want the audience to remember. Since it may be difficult for an audience to assimilate more than two or three major points in a given presentation, it is often wise to limit the talk to a few selected major points and let the presentation generate interest in written work on the topic.

The next phase of the talk is the *development of the subject*. Here the justification for the work must be established. The work of others should be referred to, and the manner in which the present study complements or supplements existing work should be described. The engineer should point out the area in which this work makes a contribution, stating appropriate justification in terms of data obtained or analysis conducted. The manner in which the work was conducted should be explained, and relative accuracy should be noted.

The engineer should next offer a statement of the findings or a *discussion of the results* of the project. In this section it would be appropriate to compare the present work with other studies and to indicate why such comparisons are favorable or unfavorable.

After the results of the project have been described, the *conclusions and recommendations* of the study should be reviewed in detail. In most instances, the recommendations will result directly from the conclusions, and that relationship should be made clear.

Finally, the *wrap-up* of the talk will restate the objectives and conclusions of the presentation. The major points you want the audience to remember should be reiterated. At this point it might be appropriate to *acknowledge* any colleagues who have assisted in the work, or agencies or organizations that may have provided support or assistance in the project.

Following the brief wrap-up, the speaker should ask whether members of the audience have *questions* about the presentation or the project. Questions from the floor are a sign of the audience's interest in the subject.

The engineer must remember to watch the time allotment for the presentation. A long, drawn-out talk or a talk which is rushed in its conclusions will be less effective than a concise, timely presentation. The speaker should move about the podium and should not read the presentation, but maintain as much eye-to-eye contact with the audience as possible. With enough practice, the speaker will become sufficiently comfortable with the content of the talk to be able to think about the audience during the presentation.

Students of public speaking have long been advised that the key to a successful presentation is to tell the audience what you are going to say, say it, and then tell them what you have said. This advice is equally valuable for engineers.

Visual aids. Visual aids are extremely important to an effective oral presentation. When well used, such devices command the attention of the audience and serve as a powerful means of communication. When poorly prepared or presented, visual aids, however, may only confuse the audience. The care with which visual aids are used is a measure of the quality of the presentation.

Many types of visual aids may be used effectively in an oral presentation: models, blackboards, large sketch pads or flip charts, overhead transparencies, slides, motion pictures, television techniques, and so forth. Each of these visual aids is appropriate for a variety of presentations.

Physical models may provide a focus for an oral presentation. Scale models, working models, or layout models are particularly useful when moving parts or various components are to be discussed during the oral presentation.

Chalkboards or *blackboards* may be used as a means of quickly communicating ideas or concepts that cannot readily be put into concise language. They are particularly useful for the presentation of equations, derivations, or rough sketches of objects or data trends.

Sketch pads or *flip charts* may be used for the presentation of ideas or salient points of a talk. These visual-aid devices are inexpensive and may be used when other equipment is not available.

The use of *transparencies* for overhead projectors requires some forethought, both in layout and in the manner in which they are used. Transparencies are easy to prepare and may provide a professional graphic display with overlays or various colored marking pens (Figure 5–18). They are particularly useful for smaller audiences.

For large audiences, 35mm or 2×2 *slides* are often the most appropriate form of visual aid, since they can be projected on a large screen for all to see. Slides provide the audience with a static picture of a process or equation. It is necessary to prepare the material for slides several days in advance of the presentation in order to permit time for the preparation of the slides themselves.

Motion pictures are ideal for conveying dynamic processes, but require preparation time as well as a great deal of expense and effort. Although the preparation of films involves significant time delays, films are useful for portraying

Figure 5–18 A contour map could be displayed for an audience by means of an overhead projection or slide. (*Courtesy* American Electric Power.)

time-dependent characteristics which may be observed through high-speed photography and shown at slower speeds.

Video tape clips are relatively easy to prepare, much like audio cassettes. Video equipment, however, is not always available. As with motion pictures, dynamic processes may be shown to the audience or instantly replayed for a better view of the action.

In preparing any visual aids for presentation, the speaker should:

(i)   Be sure that the visual aids include all the information necessary for the specific point you wish to make.

(ii) Avoid cramming too much information on a single slide or transparency.

(iii) Avoid lengthy equations or series of symbols.

(iv) Include on each slide only the salient points of the investigation or design, the major parameters or test conditions, the pertinent conclusions (including comparison with other experimental, analytical, or design studies), and the recommendations.

(v)   Allow time for the audience to absorb what you have presented.

(vi) Speak slowly and clearly so that the audience may understand the importance of each point as demonstrated by the visual aid.

When properly used, visual aids will help the engineer to make a presentation that is interesting, clear, and professional.

5.7 Summary

Communication plays a significant role in all phases of engineering. In engineering design, communication plays a unique role because of the need to convey creative ideas and designs to others for implementation. The engineer, then, must be able to effectively communicate ideas and designs through drawings, written words, and spoken words to a wide variety of audiences.

Communication is the means whereby the ideas and dreams for the future are conveyed to others. Without it, the most significant discovery, the most important innovation, will be useless. Communication is essential to success in engineering design, as it is in all phases of engineering.

EXERCISES

5.1 What do you regard as the five inventions that have most contributed to human communications?

5.2 What are the types of communications that are most often used by engineers?

5.3 Use a simple freehand sketch to illustrate the details of the following devices:

(a) adjustable wrench

(b) door-closing device

(c) two-cell flashlight

(d) pocket knife.

(e) bottle opener

(f) spray pump from a bottle of window cleaner.

5.4 What is the difference between a memorandum and a letter? Give an example of how an engineer would use each of these.

5.5 What is the difference between an engineering journal article and an engineering technical note?

5.6 What is the difference between a trade journal and a technical journal? Can you think of an example of each of these?

5.7 What is the difference between an engineering report and an engineering proposal? Give an example of how each of these might be used in engineering practice.

5.8   What types of visual aids could an engineer use in presenting a technical paper at a professional society meeting?

REFERENCES

BEAKLEY, G. C., and H. W. LEACH, *Engineering: An Introduction to a Creative Profession,* 3rd ed. New York: The Macmillan Company, 1977.

EIDE, A. R., R. D. JENISON, L. H. MASHAW, and L. L. NORTHRUP, *Engineering Fundamentals and Problem Solving.* New York: McGraw-Hill Book Company, Inc., 1979.

GLORIOSO, R. N., and F. S. HILL, JR., *Introduction to Engineering.* Englewood Cliffs, N.J.: Prentice-Hall, Inc., 1975.

GRAYSON, L. P., and J. M. BIEDENBACH, eds., *Teaching Aids in the College Classroom.* Washington, D.C.: American Society for Engineering Education, 1975.

INGLISH, J., and J. E. JACKSON, *Research and Composition, A Guide for the Beginning Researcher.* Englewood Cliffs, N.J.: Prentice-Hall, Inc., 1977.

KRICK, E. V., *An Introduction to Engineering: Methods, Concepts and Issues.* New York: John Wiley and Sons, Inc., 1976.

MANTELL, M. I., *Ethics and Professionalism in Engineering.* New York: The Macmillan Company, 1964.

RUBINSTEIN, M. F., *Patterns of Problem Solving.* Englewood Cliffs, N.J.: Prentice-Hall, Inc., 1975.

SMITH, R. J., *Engineering as a Career,* 3rd ed. New York: McGraw-Hill Book Company, Inc., 1969.

VAUGHN, R. C., *Legal Aspects of Engineering,* 3rd ed. Dubuque, Iowa: Kendall/Hunt Publishing Company, 1977.

WILSON, E. B., JR., *An Introduction to Scientific Research.* New York: McGraw-Hill Book Company, Inc., 1952.

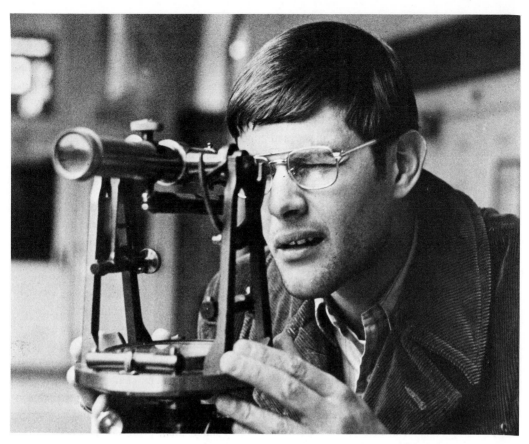

Using a transit. (*Courtesy* Luthor Gore, University of Virginia.)

Dimensions and unit systems

6

Engineering computation for design evaluation involves much more than the mere manipulation of numbers. The proper use of models and equations, the units of measurement of the physical quantities used in problems, the precision of the data used, the degree of accuracy of the answers—all these are intimately involved in the solution of problems. The basic dimensions most often used in engineering are length, mass, and time. These dimensions may be reported in a number of different units, depending upon the magnitude of the dimension involved.

The two major systems of units that have been used in recent times are the English system and the metric system. The English system of units, developed in England in the thirteenth century, has been used primarily by British Commonwealth countries and the United States. The remainder of the world has long used the metric system. In 1972, the metric system was designated the Système International d'Unités (SI) by a worldwide federation of national standards institutes including the National Bureau of Standards of the United States. This unit system is to be the universal unit system for all measurements and calculations.

The English system of units, with a few modifications, has been used extensively by American engineers. The primary modification for engineering applications was the inclusion of force units based on the gravitational system, i.e., the force due to gravity. Many engineering constants and characteristics were developed using this modified English System of units. Many practicing engineers have memorized a great many material properties in English units and have come to think in these terms. Conversion to SI units, then, poses some conceptualization difficulties. Since scientists and engineers in nearly all other countries now work in SI units, a change to this system of units by American engineers is long overdue.

As early as 1790, Thomas Jefferson recommended that the United States convert to the decimal measurement system, a system of units based on multiples of 10. His recommendation was rejected, however, because of a fear by Congress that such a conversion would interfere with U.S. trade, which was primarily with the British Commonwealth at that time. Today, however, the worldwide exchange of technical information and products that are specified in the metric system of weights and measures has led the United States to adopt the International System of Units.

6.1 The decimal system

The monetary system in the United States is based on multiples of 10. In the United States monetary system, for example, 10 mills are one penny, 10 pennies are one dime, 10 dimes are one dollar, and so forth. The system involving multiples of 10, then, is not alien to Americans, but rather is a part of everyday life.

The metric system of weights and measures is a decimal system comparable to the U.S. monetary system. A unit is 10 times larger or smaller than the next smaller or larger unit. Each of these divisions is designated by a prefix which specifies its relationship to the basic unit. These prefixes, which are listed in Table 6–1, remain the same regardless of the units involved and permit the reporting of measurements over a wide range.

Table 6–1 Metric Prefixes

SI Prefix	Symbol	Numerical Size	
tera	T	1 000 000 000 000	(10^{12})
giga	G	1 000 000 000	(10^{9})
mega	M	1 000 000	(10^{6})
kilo	k	1 000	(10^{3})
hecto	h	100	(10^{2})
deka	da	10	(10^{1})
deci	d	0.1	(10^{-1})
centi	c	0.01	(10^{-2})
milli	m	0.001	(10^{-3})
micro	μ	0.000001	(10^{-6})
nano	n	0.000000001	(10^{-9})
pico	p	0.000000000001	(10^{-12})
femto	f	0.000000000000001	(10^{-15})
atto	a	0.000000000000000001	(10^{-18})

6.2 Fundamental dimensions

In the International System of Units (SI), there are seven primary or fundamental dimensions of physical quantities from which other units are derived. These fundamental quantities and recommended symbols are given in Table 6–2. These

Table 6–2 Fundamental Units—International System of Units (SI)

Primary Unit	Symbol	Unit of Measure	Abbreviation
Length or distance	*1*	meter	m
Mass	*m*	kilogram	kg
Time	*t*	second	s
Thermodynamic temperature	*T*	kelvin	K
Electric current	*I*	ampere	A
Amount of substance	*n*	mole	mol
Light or luminous intensity	l_v	candela	cd
Supplementary Units			
Plane angle	$\alpha, \beta,\ \gamma, \theta, \phi$	radian	rad
Solid angle	ω	steradian	sr

basic quantities are defined in terms of the standards (criteria established by authority or agreement as the measure of a quantity) used to establish the primary units.

Length or distance. The measurement of the length of or distance to an object, such as the length or diameter of a piece of steel rod, is reported in terms of *meters*. The meter is defined as being exactly "the length equal to 1 650 763.73 wavelengths in vacuum of the radiation corresponding to the transition between the levels $2p_{10}$ and $5d_5$ of the krypton-86 atom"* (see Figure 6–1).

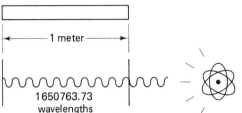

1 meter

1 650 763.73
wavelengths

Figure 6–1 Schematic of standard for length.

Normal distances or measurements of length may be reported in terms ranging from micrometers (10^{-6} meters) to kilometers (10^3 meters) and beyond. Conversion between these units is easily accomplished since they are related by factors of 10. Manipulation of areas (square meters) or volumes (cubic meters) in terms of SI units minimizes the number of steps in computations involving these units.

Mass. It should be clearly understood that the measurement of the mass of an object, such as a brick, is not the same as the measurement of weight. The mass of the brick is a constant, whereas the weight depends upon gravitational forces which vary with such conditions as altitude and location. The weight of the brick, then, is the force with which the brick is pulled downward (Figure 6–2).

The *kilogram* is the unit used to describe mass (not weight or force). In common parlance, the word "weight" often is used to refer to what is actually the "mass" of an object. These two values are identical, however, only at standard sea level conditions. The standard for the kilogram is a cylinder of platinum-

*National Bureau of Standards, SP 330, p. 3.

Figure 6–2   Zeroing the scale before weighing an object.
(*Courtesy* Luther Gore, University of Virginia.)

iridium alloy maintained under vacuum conditions by the International Bureau of Weights and Measures in Paris (Figure 6–3). A duplicate of this cylinder of

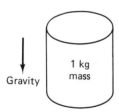

Figure 6–3   Schematic of the standard for mass.

platinum-iridium alloy is kept by the National Bureau of Standards in Washington, D.C. Mass is the only primary unit for which a tangible object is maintained as a standard.

The relationship between force and mass is defined by means of Newton's second principle:

$$\text{Force} = (\text{mass})\,(\text{acceleration})$$

The force unit is defined in terms of the units of mass, length, and time. The absolute force unit in the International System is the *newton*, designated by the symbol N, and is the force required to accelerate one kilogram of mass one meter per second per second, as shown in Figure 6–4, that is:

$$1\ N = 1\ \text{kg m/sec}^2$$

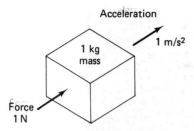

Figure 6–4 Schematic of the relationship between force and mass.

Time. Time is the duration between two distinct events, or the elapsed period since an event has occurred. The basic unit of time is the *second,* which is defined as exactly "the duration of 9 192 631 770 periods of the radiation corresponding to the transition between the two hyperfine levels of the ground state of the cesium-113 atom."* The second is determined by turning an oscillator to the resonant frequency of cesium-113 atoms passing through a magnetic field in a resonant cavity, sometimes referred to as the cesium clock.

There are secondary time standards, in addition to the cesium clock, which may be used for defining the second. Among these are the hydrogen maser, rubidium clock, and quartz frequency standards and clocks.

Time is measured in terms of seconds, minutes, hours, days, weeks, months, and years. For periods of time of one second or less, the decimal system is used, e.g., milliseconds or microseconds. (Five milliseconds may be written 0.005 seconds; 27 microseconds may be written 0.000 027 seconds.) For periods of time from one second to years, individual increments are involved and do not follow the decimal system. There are 60 seconds in a minute (rather than 100), 60 minutes in an hour, 24 hours in a day, 7 days in a week, and so on. For periods greater than one year, the decimal system is again used, i.e., 10 years is a decade, 10 decades a century, and 10 centuries a millennium.

Electric current. Electric current is the rate of flow of electric charge through an electrically conducting material. The unit of electric charge is the coulomb. The charge on an electron has been observed in measurements on the hydrogen atom. This charge, designated by the symbol e, is 1.6021×10^{-19} coulombs.

The flow of electric charge through an electrically conducting material, such as a length of wire, produces a force field. If an electric current is passed through two long parallel wires separated by one meter of free space, a magnetic field will be produced between the wires. The *ampere* is defined as "that constant current which, if maintained in two straight parallel conductors of infinite length, of neg-

*NBS SP 330, p. 3.

ligible circular cross section, and placed one meter apart in a vacuum, would produce between these conductors a force equal to 2×10^{-7} newtons per meter of length''* (see Figure 6–5). The ampere is a measure of the rate of flow of charge and is one coulomb per second. It should be noted that the flow of electrons through a conducting medium is opposite to the direction of current.

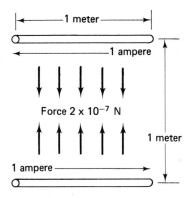

Figure 6–5 Schematic of the standard for the ampere.

Temperature. The measure of temperature is a measure of relative ''hotness'' or ''coldness.'' In 1854 Lord Kelvin proposed a thermodynamic temperature scale based on six fixed temperature points including the freezing and boiling points of water. In 1968 the International Practical Temperature Scale was adopted by an international conference concerned with the establishment of temperature standards. This International Temperature Scale is based on a number of fixed points, as noted in Table 6–3.

Several different instruments are used to measure temperatures for use in standard temperature measurements. Among these instruments are the platinum resistance thermometer, the platinum-platinum/10% rhodium thermocouple, and the monochromatic optical pyrometer. These instruments are calibrated in terms of the reproducible temperature points, or fixed points, some of which are listed in Table 6–3.

The *kelvin* is used as a measure of temperature in the International System of Units and has been defined as ''the fraction $^{1}/_{273.16}$ of the thermodynamic temperature of the triple point of water,''* that is, the point at which water exists in all three phases—solid, liquid, and vapor. A special calibration cell has been constructed which will permit the calibration of temperature measuring devices at the

*NBS SP 330, p. 4.

Table 6–3 Temperature Scales

| International System | | | | English System | |
Kelvin	Celsius	Standard		Rankine	Fahrenheit
K	°C			°R	°F
0.0	−273.15	Absolute zero		0.0	−459.69
20.28	−252.87	Hydrogen liquid-vapor equilibrium		36.51	−423.19
90.188	−182.962	Oxygen liquid-vapor equilibrium		162.344	−297.346
273.15	0.00	Water solid-liquid equilibrium		491.69	32.00
273.16	0.010	Water triple point		491.71	32.018
373.15	100.0	Water liquid-vapor equilibrium		671.68	212.00
692.73	419.58	Zinc solid-liquid equilibrium		1 246.93	787.24
717.82	444.67	Sulfur liquid-vapor equilibrium		1 291.97	832.28
1 235.08	961.93	Silver solid-liquid equilibrium		2 221.09	1 761.4
1 337.58	1 064.43	Gold solid-liquid equilibrium		2 405.09	1 945.4
1 827	1 554	Palladium solid-liquid equilibrium		3 288	2 829
2 044.8	1 771.7	Platinum solid-liquid equilibrium		3 680.8	3 221.1

Scale markings (left margin): 0, 100, 200, 300, 400, 500, 600, 700, 800, 1200, 1300, 1800, 1900, 2000, 2100

triple point. The kelvin scale, generally referred to as the "absolute temperature scale," is frequently used by scientists and engineers.

The Celsius scale, usually referred to as the Centigrade scale (i.e., 100 graduations between freezing and boiling of water) also is used in the International System of Units. Although the increment of one degree Celsius is equal to the increment of one degree kelvin, the magnitude of a given temperature in degrees Celsius is different from the magnitude in degrees kelvin, as noted in Table 6–3. Absolute zero is 0 K and $-273.15°C$, and 273.15 K equals 0°C. Therefore $°C = K - 273.15$. The kelvin scale and the Celsius scale, then, are closely related and both are a part of the International System of Units.

Amount of substance.   In many analyses, there is a need to refer to quantities of substance or amount of constituent particles (Figure 6–6). In the past, terms involving the number of atoms or molecules were used to specify amounts of chemical elements or chemical compounds. These terms were directly related to the molecular or atomic weights of the substance. Isotopes of oxygen were used as a standard for a period of time. In 1960, an agreement was reached to establish a unified scale of relative atomic mass with the isotope Carbon 12 as the base.

The unit of amount of a substance was established as the *mole,* and has been fixed as "the amount of a substance which contains as many elementary entities

Figure 6–6 Engineer measures the amount of a chemical compound. (*Courtesy* General Electric Company.)

as there are atoms in 0.012 kilograms of Carbon 12.''* These elementary entities may be atoms, molecules, ions, electrons, or other particles or groups of particles.

Consider a pure substance (chemical symbol A) composed of atoms. By definition, a mole of atoms of A contains as many atoms as there are Carbon 12 atoms in 0.012 kilograms of Carbon 12. Since the masses of these atoms in both substance A and Carbon 12 cannot readily be measured, a ratio of the masses can be determined, such that

$$M(A) = \left(\frac{m(A)}{m(^{12}C)} \right) \times 0.012 \text{ kg/mol}$$

where M is the mass of a mole of substance A. The ratio may be determined by mass spectrograph. It is possible, then, to calculate the molar mass of any substance by utilizing the ratio of the mass of any substance to the mass of Carbon 12.

A mole of particles of any perfect gas at a given temperature and pressure occupies the same volume. Therefore, it is possible to determine the ratio of amounts of substance for gases. For electrolytic solutions (i.e., solutions in which the conduction of electricity is accompanied by chemical decomposition), the ratio of amounts of substance is proportional to the electrical output. Other techniques may be used to establish a ratio of the amounts of substance for extremely dilute solutions.

Light or luminous intensity.   Virtually all bodies emit or absorb radiant energy. If the temperature of a body is sufficiently high, the emitted radiation will be in the visible spectrum. Thermal radiation is not the only form of light emission. Other forms include fluorescence, chemiluminescence, electrical excitation, and electroluminescence. In order to classify the degree of visible light emitted, a unit of measure has been adopted in the International System. This unit of measure of luminous intensity is the *candela,* and is defined as ''the luminous intensity, in the perpendicular direction, of a surface of 1/600 000 square meters of a blackbody at the temperature of freezing platinum under a pressure of 101 325 newtons per square meter''* (see Figure 6–7).

There are, in addition, secondary standards which may be used for measurement of luminous intensity. Among these are specially manufactured incandes-

*NBS SP 330, p. 5.

Blackbody

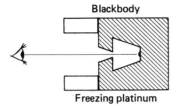

Freezing platinum

Figure 6–7 Schematic of the Standard for the Candela.

cent lamps with d.c. power supplies. These lamps are calibrated to provide definite luminous intensities at given voltage settings (Figure 6–8).

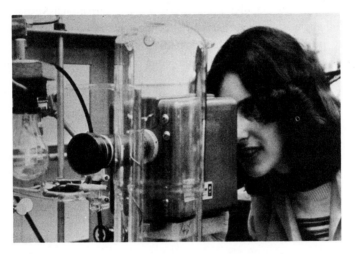

Figure 6–8 Measurement of the luminous intensity of a lightbulb. (*Courtesy* General Electric Company.)

Plane angle. One of the more commonly used supplementary units of angular measure is the plane angle. This angle is designated as the *radian* and is defined as "the plane angle between two radii of a circle which cut off on the circumference an arc equal in length to the radius"* (see Figure 6–9). There are 2π radians in a circle, and therefore in 360°. As a result, one radian is 57.2957795°. In addition,

$$1 \text{ minute } (1') = (1/60)° = \pi/10\ 800 \text{ radians; and}$$

$$1 \text{ second } (1'') = (1/60)' = \pi/648\ 800 \text{ radians}$$

*NBS SP 330, p. 11.

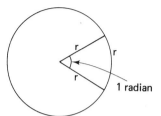

Figure 6–9   Schematic of the radian.

Solid angle.   The solid angle is used in several types of engineering calculations and is one of the supplementary units of the International System. The measure of solid angle is the *steradian* and is defined as "the solid angle which, having its vertex in the center of a sphere, cuts off an area of the surface of sphere equal to that of a square with sides of length equal to the radius of the sphere"* (see Figure 6–10).

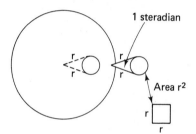

Figure 6–10 Schematic of the Steradian.

6.3 Derived units

The previously described primary units may be used to formulate more complex units for specific situations. These additional units, or derived units, are formed algebraically by multiplication or division (Figure 6–11). Some of these derived units have been given special names and symbols. A number of these units are listed in Table 6–4.

The engineer often is called upon to work with both the International System of Units and the English system. To assist in this process, a detailed table of conversion factors is given in Appendix B.

*NBS SP 330, p. 11.

Figure 6–11 Area is a derived unit which may be measured with a planimeter. (*Courtesy* Luther Gore, University of Virginia.)

Table 6–4 SI Basic and Derived Units*

Quantity	Symbol	Unit of Measure	Abbreviation
Basic			
length	*l*	meter	m
mass	*m*	kilogram	kg
time	*t*	second	s
thermodynamic temperature	*T*	kelvin	K
electric current	*I*	ampere	A
amount of substance	*n*	mole	mol
luminous intensity	l_v	candela	cd
plane angle	$\alpha, \beta, \gamma, \theta, \phi$	radian	rad
solid angle	ω	steradian	sr
Derived			
acceleration		meter/second squared	m/s²
activity (of radioactive source)		disintegration/second	s^{-1}
angular acceleration		radian/second squared	rad/s²

Table 6–4 SI Basic and Derived Units* (continued)

Quantity	Symbol	Unit of Measure	Abbreviation
angular velocity		radian/second	rad/s
area		square meter	m²
concentration		mole/cubic meter	mol/m³
current density		ampere/square meter	A/m²
density (mass)		kilogram/cubic meter	kg/m³
electric capacitance		farad (F)	A·s/V
electric conductance		siemens (S)	A/V
electric charge		coulomb (C)	A·s
electric field strength		volt/meter	V/m
electric inductance		henry (H)	V·s/A
electric potential difference		volt (V)	W/A
electric resistance		ohm (Ω)	V/A
electromotive force		volt (V)	W/A
energy		joule (J)	N·m
energy density		joule/cubic meter	J/m³
entropy		joule/kelvin	J/K
force		newton (N)	kg·m/s²
frequency		hertz (Hz)	cycle/s
heat		joule (J)	N·m
heat capacity		joule/kelvin	J/K
heat flux density		watt/square meter	W/m²
illuminance		lux (lx)	lm/m²
luminance		candela/square meter	cd/m²
luminous flux		lumen (lm)	cd·sr
magnetic field strength		ampere/meter	A/m
magnetic flux		weber (Wb)	V·s
magnetic flux density		tesla (T)	Wb/m²
magnetomotive force		ampere (A)	C/s
permeability		henry/meter	H/m
permittivity		farad/meter	F/m
power		watt (W)	J/s
pressure		pascal (Pa)	N/m²
quantity of electricity		coulomb (C)	A·s
quantity of heat		joule (J)	N·m
radiance		watt/square meter steradian	W/m²sr
radiant intensity		watt/steradian	W/sr
radiant flux		watt (W)	J/s
specific energy		joule/kilogram	J/kg
specific heat		joule/kilogram·kelvin	J/kg·K

Table 6–4 SI Basic and Derived Units* (continued)

Quantity	Symbol	Unit of Measure	Abbreviation
specific volume		cubic meter/kilogram	m³/kg
speed		meter/second	m/s
stress		pascal (Pa)	N/m²
surface tension		newton/meter	N/m
thermal conductivity		watt/meter·kelvin	W/m·K
thermal conductance		watt/square meter kelvin	W/m²K
velocity		meter/second	m/s
viscosity, dynamic		pascal·second	Pa·s
viscosity, kinematic		square meter/second	m²/s
voltage		volt (V)	W/A
volume		cubic meter	m³
wave number		reciprocal meter	m⁻¹
work		joule (J)	N·m

*NBS SP 330.

6.4 Checking units

Once the parameters for a problem have been established, it is necessary to check all the units involved in order to be sure that they are specified in the same unit system (i.e., that they are homogeneous), and that the units of the answer will be in the desired form. If data with different units are multiplied or added together, erroneous answers will result which may become the basis for erroneous judgments. The combination of apples and oranges to obtain an "orple" would hardly be a fruitful exercise.

In preparing a problem solution, the fundamental equations used also must be dimensionally correct. Since an equation may involve several variables, the test of dimensional homogeneity can be made by substituting the appropriate units for each variable in the equation (without numbers). The units on the left side of the equation should be equal to those on the right side.

There are occasions when the problem solver would like to reduce the calculation time for a specific problem by rearranging the problem into groups of dimensionless quantities. One of the more frequently used procedures in engineering is the dimensional analysis technique. This technique permits the combination of several variables into one dimensionless quantity for easier calculation. Basically, these dimensionless quantities are the result of multiplication or division of terms involving the basic units. In checking the units of such dimensionless numbers, the individual units of the numerator and denominator should be equal.

EXAMPLE 6–1   Kinetic energy is defined as

$$KE = \tfrac{1}{2} MV^2$$

Using appropriate units from Table 6–4:
      Energy is measured in terms of joules
      Mass .... kg
      Velocity .... m/s
Then substituting in the kinetic energy equation,

$$\text{joule} = \text{kg}\frac{\text{m}^2}{\text{s}^2}$$

From Table 6–4, the joule is defined as

$$\text{joule} = \text{kg m}^2 \text{ s}^{-2}$$

The equation, therefore, is dimensionally correct.

EXAMPLE 6–2   The drag coefficient used in aerodynamics is defined as

$$C_d = \frac{D/A}{\rho \, (v^2/2)}$$

where: D = drag force
       A = area over which drag force acts
       ρ = density
       v = velocity

In order to demonstrate that C_d is a dimensionless number, from Table 6–4 we find

Drag force is measured in terms of newtons
Area .... m²
Density .... kg/m³
Velocity .... m/s

Substitution of these values into the drag coefficient equation yields:

$$C_d = \frac{N/m^2}{(kg/m^3)\,(m^2/s^2)} = \frac{N}{N} = 1$$

Thus C_d is dimensionless.

There are numerous references at the end of this chapter describing dimensional analysis techniques, including the Buckingham theorem (a technique for the selection of dimensionless groups of variables), for those interested in the reduction of variables.

Unit checking of equations not only clarifies the information needed to solve the problem, but also assures that the correct units and conversion factors are involved. It is an essential part of engineering computations.

6.5 Summary

A substantial effort has been made in recent years to standardize the various units and dimensions used in science and engineering. The adoption of the International System of Units simplifies the international exchange of information and products, and provides new opportunities for collaboration on scientific problems of worldwide scope.

Once the engineer has developed an understanding of the basic SI units and has used them repeatedly in problem solving, he or she will develop a perception of the appropriate magnitudes of various parameters and their use in engineering computations. Knowledge of SI units and their use, then, will facilitate the solution of engineering problems.

PROBLEMS **6.1** Energy is generated when a current is passed through a resistance. Show that the equation

$$P = I^2R$$

is dimensionally homogeneous if

$$P = \text{power}$$

$$I = \text{current}$$

$$R = \text{electric resistance}$$

6.2 The speed of sound in an enclosure filled with gas may be calculated as

$$C = (kRT/M)^{1/2}$$

where: C = speed

R = universal gas constant

T = thermodynamic temperature

k = ratio of the specific heat at constant pressure to the specific heat at constant volume

M = molecular mass of the gas

What should the units of R be in order for the equation to be dimensionally homogeneous?

6.3 When a conductor moves through a magnetic field, the voltage, v, generated may be determined by the expression

$$V = BLv$$

where: B = magnetic flux density

L = length of the conductor

v = velocity of the conductor through the field

Show that the equation is dimensionally homogeneous.

6.4 The normal modes of vibration in a string fixed at each end result in frequencies expressed by the equation

$$f_n = \frac{n(F/m)^{1/2}}{2L}$$

where: f = frequency

F = tension

m = mass/unit length

L = length of string

n = mode of vibration

Show that the equation is dimensionally homogeneous.

6.5 When a current-carrying conductor is placed in a magnetic field, a force, F, is generated according to the equation

$$F = BLI$$

where: B = magnetic flux density

L = length of the conductor

I = current

Show that the equation is dimensionally homogeneous.

6.6 In the field of optics, the equation for a double convex lens may be written as

$$\frac{1}{l_1} + \frac{1}{l_2} = (\mu - 1)\left(\frac{1}{R_1} + \frac{1}{R_2}\right)$$

where: l_1 = object distance

l_2 = image distance

R = radius of curvature of lens surfaces

μ = index of refraction of lens material

What are the units of μ?

6.7   Determine the units for the variable K in the Euler formula

$$\frac{P}{A} = \frac{K\,\pi^2\,E}{(L/R)^2}$$

where: A = area

E = modulus of elasticity

R = radius of gyration

L = column length

P = force

6.8   If an electron of mass m is permitted to fall through a voltage, V, the velocity, v, reached by the electron may be determined through use of the equation

$$v = (2\ eV/m)^{1/2}$$

where: e = electric charge

V = voltage

m = mass

v = velocity

Show that the equation is dimensionally homogeneous.

6.9   The Bernoulli equation may be used for a nonviscous, incompressible steady flow through a pipe of variable cross section. The equation between points 1 and 2 may be written as

$$p_1 + \tfrac{1}{2}\rho v_1^2 + \rho g z_1 = p_2 + \tfrac{1}{2}\rho v_2^2 + \rho g z_2$$

where: p = pressure

ρ = density

v = velocity

g = gravitational acceleration

z = distance

Show that the equation is dimensionally homogeneous.

6.10 Determine the units for dimensional homogeneity in the equation for the Stormer type rotating viscometer.

$$\mu = \frac{\Delta\Theta(r_2{}^2 - r_1{}^2)r_3{}^2 \; g \; \Delta m}{4\pi r_1{}^2 r_2{}^2 z(h+\delta)}$$

where: $r_1, r_2, r_3, z, h, \delta$ = units of length

g = acceleration due to gravity

Θ = time $\Delta\Theta$ = increment of time

m = mass Δm = increment of mass

6.11 Demonstrate that the following dimensionless parameters are without units. (Use the information given in Table 6–4.)

(a) Reynolds Number, $Re = \rho Vd/\mu$

ρ = density

V = velocity

d = diameter

μ = viscosity

(b) Nusselt Number, $Nu = hD/k$

h = convective conductance

D = length

k = thermal conductivity

(c)   Prandtl Number, $Pr = \mu c_p / k$

  μ = viscosity

  c_p = specific heat

  k = thermal conductivity

(d)   Biot Number, $Bi = hr/k$

  h = convective conductance

  r = radius

  k = thermal conductivity

(e)   Froude Number, $Fr = V^2/Lg$

  V = velocity

  L = length

  g = acceleration due to gravity

6.12   A monkey has a mass of 10 kg. What is the force exerted by the monkey (in newtons) at standard sea level conditions near the equator where g is 9.82 m/s^2, and in a satellite orbiting 400 kilometers from earth where g = 9.0 m/s^2? What would the weight of the monkey be at the same conditions?

6.13   Calculate the mass of an inverted conical container filled with water if the altitude of the conical section is 43 cm and the radius of the base is 8 in. (Assume standard conditions.)

6.14   The total distance traveled by an automobile at constant acceleration is

$$s = v_0 t + \tfrac{1}{2}a\,t^2$$

and the final velocity reached is

$$v_f = (v_0^2 + 2a\,s)^{1/2}$$

Calculate the distance, s (in km traveled), and final velocity, v_f (in km/hr), if

$$v_0 = 10 \text{ miles/hour}$$
$$t = 60 \text{ sec}$$
$$a = 4 \text{ ft/sec/sec}$$

6.15 With what force (in newtons) will a balloon filled with water hit a designated target when it is dropped from a third story window 10 m above the given target. Assume that the balloon is a rigid sphere 12 in. in diameter. (Use standard conditions.)

6.16 Calculate the heat loss through a brick furnace wall (in joules/m²) using the Fourier conduction equation

$$Q = \frac{kA(T_2 - T_1)}{L}$$

where: $T_2 = 2000°F$ (inside temperature)

   $T_1 = 100°C$ (outside temperature)

   $k = 0.25$ Btu/hr/ft²/°F/ft (thermal conductivity)

   $L = 23$ cm (furnace wall thickness)

   $A = 1$ m² (surface area)

6.17 Calculate the kinetic energy (in newton-meters) for a bowling ball rolling down a lane in a bowling alley if the ball weighs 5 lb and the average velocity is 700 cm/s. (Assume standard conditions.)

6.18 The fire hydrants in a city water system are being tested. Calculate the velocity, v, of the water (in m/s) flowing out of the hydrant if the center of the outlet is 75 cm above the ground and the issuing jet of water strikes the ground 9 ft from the outlet. The velocity may be calculated from

$$v = \sqrt{\frac{x^2 g}{2y}}$$

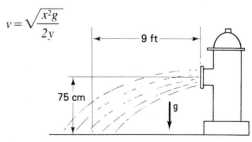

(Assume standard conditions.)

6.19 The deflection, δ, of a cantilevered flat parallel spring is given by the expression

$$\delta = \frac{2}{3} \frac{sL^2}{Et}$$

where the safe tensile stress, s, is 50,000 psi, the modulus of elasticity, E, is 30×10^6 psi for steel springs, the length, L, is 0.75 m, and the thickness, t, is 0.5 cm. Calculate the deflection of the spring in cm.

6.20 The general equation for pressure loss due to pipe friction is

$$\Delta p = fp \frac{L}{D} \frac{v^2}{2}$$

where: Δp = pressure loss, newtons/m²
f = friction factor, dimensionless
ρ = density of fluid flowing in pipe, kg/m³
L = pipe length, m
D = pipe diameter, m
v = velocity, m/s
g = gravitational acceleration, m/s²

Water is flowing through a 2-in. diameter pipe, 20 ft long, at 200 ft/min. Calculate the pressure loss if the friction factor is 0.02.

6.21 A new internal combustion engine is installed on a dynamometer to measure the engine's performance. The power, P, is determined by the relationship

$$P = \frac{2\pi NFL}{C}$$

where the constant, C, is 33,000 ft lb/hp min. The force, F, exerted on the 40-cm moment arm, L, is 25 kg when N is 3000 revolutions per minute. Calculate the power in watts.

6.22 Calculate the Reynolds number for air at atmospheric pressure flowing in a pipe with an inside diameter of 2 in., if the viscosity of the air is 0.021 centipoise, the density is 0.075 lb/ft³, and the velocity is 1300 cm/s. Reynolds number is defined as

$$\text{Reynolds Number} = \text{Re} = \frac{\rho D v}{\mu}$$

where: ρ = density
D = pipe diameter
v = velocity
μ = viscosity

6.23 Einstein's theory of relativity may be expressed as

$$E = mc^2$$

where E is energy, m is mass, and c is the speed of light. One of the consequences of this theory is that mass may be converted to energy. Calculate the energy equivalent of one kilogram of mass.

6.24 Newton's law of gravitation may be expressed as

$$F = G\frac{m_1 m_2}{r^2}$$

where F is the force of attraction between two bodies of masses m_1 and m_2 separated by a distance r, and G is the gravitational constant. If m_1 is 50 gm, m_2 is 125 gm, and r is 2.73 m, calculate the force between the two masses.

6.25 Einstein's theory of relativity demonstrates that mass increases with speed, that is

$$m = m_0/[1 - (v/c)^2]^{1/2}$$

where m_0 is the rest mass or mass with respect to the observer, v is the velocity, and c is the speed of light. If a neutron is traveling at $0.85c$ and the rest mass is 1.67474×10^{-24} gm, what is the mass of the particle?

REFERENCES

ACKLEY, R. A., *Physical Measurements and the International (SI) System of Units*, 3rd ed. San Diego: Technical Publications, 1970.

"AISI Metric Practice Guide: SI Units and Conversion Factors for the Steel Industry." Washington: American Iron and Steel Institute, 1975.

BURTON, W. K., ed., "Measuring Systems and Standards Organizations." New York: American National Standards Institute, Inc., 1970.

CHISWELL, B., and E. C. M. GRIGGS, *SI Units*. Sydney: John Wiley & Sons, 1971.

"Conversion to the Metric System of Weights and Measures." Hearings before the Subcommittee on Science, Research, and Technology of the Committee on Science and Technology, U.S. House of Representatives. Washington: U.S. Government Printing Office, 1975.

DE SIMONE, D. V., ed., *A Metric America: A Decision Whose Time Has Come*. National Bureau of Standards Special Publication 345, July 1971.

EIDE, A. R., R. D. JENISON, L. H. MASHAW, and L. L. NORTHUP, *Engineering Fundamentals and Problem Solving*. New York: McGraw Hill Book Company, Inc., 1979.

FLETCHER, L. S., and T. E. SHOUP, *Introduction to Engineering Including FORTRAN Programming*. Englewood Cliffs, N.J.: Prentice-Hall, Inc., 1978.

MECHTLY, E. A., "The International System of Units: Physical Constants and Conversion Factors," Second Revision. NASA SP-7012, 1973.

PAGE, CHESTER H. and PAUL VIGOUREUX, eds., "The International System of Units (SI)." National Bureau of Standards Special Publication 330, July 1974.

"SI Units and Recommendations for the Use of Their Multiples and of Certain Other Units." International Organization for Standardization, ISO 100, 1973.

"Standard for Metric Practice," E 380-79, American Society for Testing and Materials, 1980.

STIMSON, H. F.: "The International Temperature Scale of 1948," *NBS J. Research*, vol. 42, March 1949.

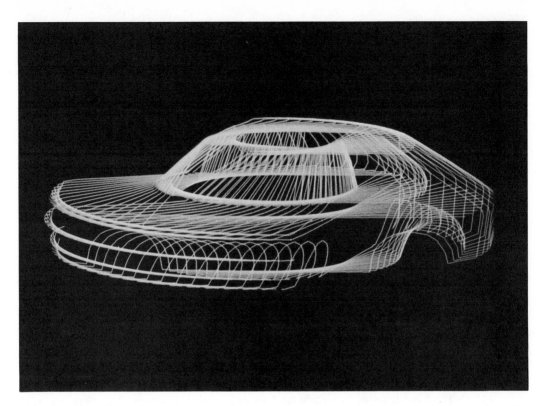

Computer-aided design. (*Courtesy* Chrysler.)

Engineering graphics

7

Many ideas and concepts in design can be expressed more clearly, more exactly, and more concisely by using graphic language in place of or in addition to words. In the initial stages of design, the designer may test out his or her own ideas by means of rough sketches. As these ideas develop, a careful scale drawing may be made to check sizes of parts which must all fit together to function properly. When the final design is ready for manufacturing, a series of detailed drawings is needed, giving a complete description of every part to be produced. The customer who purchases the finished product may receive a drawing which describes how the design can be disassembled for repair purposes. Drawings might also be furnished to the sales force for describing the product to potential customers. Engineering drawings play a vital part in virtually every phase of the design process.

Many components are combined when creating an engineering drawing. Lettering, lines, a variety of views, and numerous conventions are selected to convey some particular information to the reader of the drawing. These components are described in detail in the remainder of this chapter.

7.1 Lettering

Freehand lettering is widely used on engineering drawings. The most important test of lettering of any kind is its legibility. If lettering is of such poor quality that an error occurs in reading it, disastrous results may occur. Lettering cannot be learned simply by reading this book. Patience and practice are necessary to the development of skill in lettering. This book can, however, provide some samples of good lettering: lettering which is simple, clear, and relatively easy to produce.

Figures 7–1 and 7–2 show the letters and numbers in single stroke Gothic, the style widely used on engineering drawings.

In Figure 7–1, the characters are grouped by proportions. Twenty-six alphanumeric characters use a 6×7 grid, and eight others use a square 7×7 grid. Two characters, the "one" and the "i", are simply vertical lines.

Figure 7–2 shows suggested capital and lower case characters in single-stroke Gothic for use on engineering drawings. It is important to note that most of the letters are almost as wide as they are high.

Lettering produced on plotters as computer output is generally very similar to the single-stroke Gothic style. A common difference is the attempt by plotter characters to differentiate between "one" and "I", and "zero" and the letter "O", as shown in Table 7–1. Eventually the plotter style is likely to be incorporated into freehand styles in order to avoid possible reading errors. Another style of simple, legible lettering used in the business community is the microfont letter system shown in Figure 7–3.

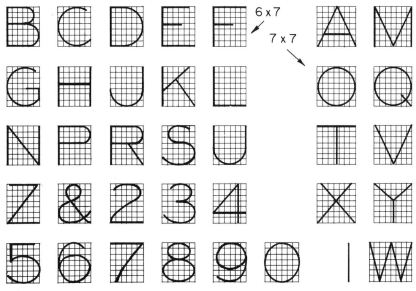

6 x 7

7 x 7

8 x 7

Figure 7–1 Proportions of vertical capital single stroke Gothic letters.

Figure 7–2 Vertical capital and lowercase letters. (American National Standards Institute [ANSI] 14.2 M–1979.)

Table 7–1 Comparison of Lettering Styles

	I	One	Zero	Letter "O"
Freehand	I	I	0	O
Plotter	I	1	0	O

ABCDEFGHIJKLMNO

PQRSTUVWXYZ

1234567890

Figure 7–3 Microfont letters. (ANSI 14.2 M–1973.)

In order to reproduce letters of these types, certain planning is required. Before lettering on plain paper, guidelines should be drawn in very lightly. Figure 7–4 shows the two horizontal guidelines normally used for capital letters and the four guidelines used for lower-case letters. The lower-case letters are generally two-thirds the height of the capital letters, with another third on some letters extending either above or below this height. An occasional vertical line will serve as a guide for keeping letters uniformly vertical. After light guidelines are drawn, the words are formed using a space equal to an ''O'' between words and less than half this space between letters within a word. Examples of good and poor spacing are shown in the lower part of Figure 7–4. The area between letters should be approximately equal for a pleasant appearance. This suggestion does not mean equal spacing. The shape of the letters determines the distance between them. For example, T and A can overlap and still allow ample area, but M and E need extra distance between them.

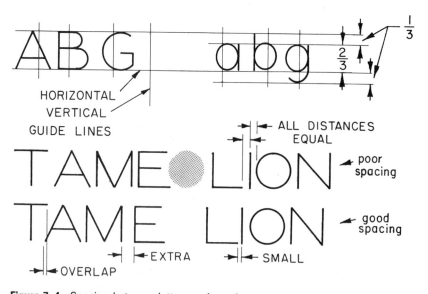

Figure 7–4 Spacing between letters and words.

Although freehand lettering is adequate for many engineering drawings, greater precision may be achieved through use of one of the many mechanical aids currently available for lettering drawings. A popular device for producing lettering on ink drawings is the Leroy lettering set, shown in Figure 7–5. This set includes a penholder, a pen with a supply of ink, and a template which determines the size and shapes of letters.

Figure 7–5 Leroy lettering equipment. (*Courtesy* K&E.)

7.2 Sketching

Freehand sketching is an important drawing skill which requires very little equipment, usually only pencil, paper, and eraser. Some sketches are done by the designer to test out preliminary ideas and may never be seen by others. Other sketches may be used to furnish draftsmen information necessary to produce final drawings for fabrication.

Technical sketching requires both skill and knowledge of common drawing conventions and practice. This textbook can illustrate the various rules or conventions used in sketching and can suggest means of improving skills, but practice is essential in learning to sketch.

Conventions in sketching involve choosing the correct types of lines, orienting the views, and selecting the types of views appropriate to the project. For example, the three-dimensional object shown in Figure 7–6 can be sketched in a number of different *pictorial views*. Note that the object has been placed in many different positions, some showing the surface labeled "A", others not. These include isometric (a), oblique (g), and perspective (e) views. In order to sketch each of these three different types of pictorial views, a particular set of rules must be followed. All these types of sketches, however, have many traits in common. Pictorial views show only the portions of the object visible for the particular direction from which it is viewed.

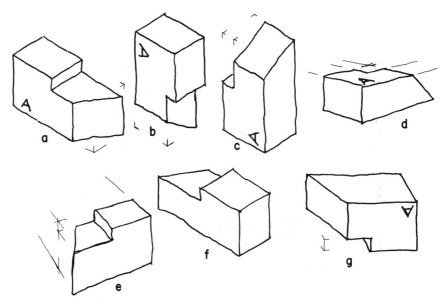

Figure 7–6 Seven pictorial sketches of the same object.

A *multiview sketch* is frequently used to depict a three-dimensional object. For this type of drawing, the object is assumed to be inside a glass box, with one particular face of the object parallel to the front of the box. In Figure 7–7, face "A" has been selected as the front of the object and is parallel to the front of the box. The object can be positioned in any way the sketcher chooses, but this decision as to orientation should be made before the sketch is begun.

For a multiview sketch, the top, front, and side views are drawn in the positions shown in Figure 7–8. The explanation for this arrangement of views is relatively simple: first, the object is viewed from the top, front, and side, and the

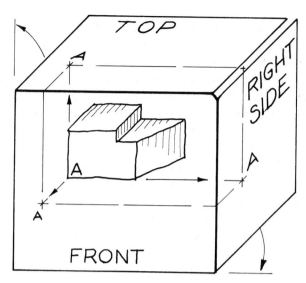

Figure 7–7 Projection of a point in space on three mutually perpendicular planes (orthographic projection.)

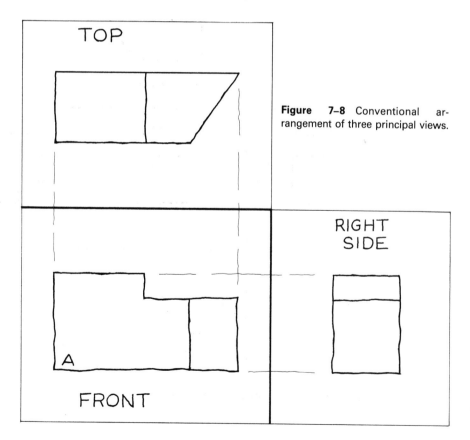

Figure 7–8 Conventional arrangement of three principal views.

three views are projected on these three faces of the box. Next, the top and side faces of the box are rotated 90° so they coincide with the front face of the box, forming a single plane, and this plane coincides with the paper on which the sketch is made. The top view, then, is always sketched above the front view, and the side views are sketched to the side of the front view. If the box is also considered to have a bottom and a left side, the five faces would then appear as shown in Figure 7–9. One of the main reasons for sketching multiview drawings in these relative positions is that dimensions are shared by adjacent views, reducing the number of dimensions which must be estimated.

Pictorial views (Figure 7–6) and multiviews (Figure 7–9) differ by the fact that multiviews show all features, visible and hidden, while the pictorials generally only show visible features. In multiviews, hidden features are shown as dashed lines (– – – –) to distinguish them from visible solid lines (————). The center line is used in multiview drawings when circular shapes occur. In Figure 7–10, the cylinder and the cylindrical hole have center lines in both the top and

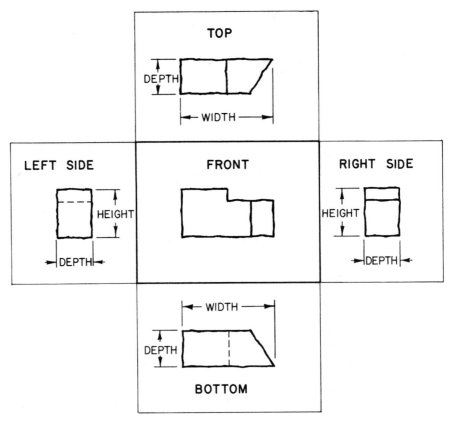

Figure 7–9 Five principal views in conventional arrangement.

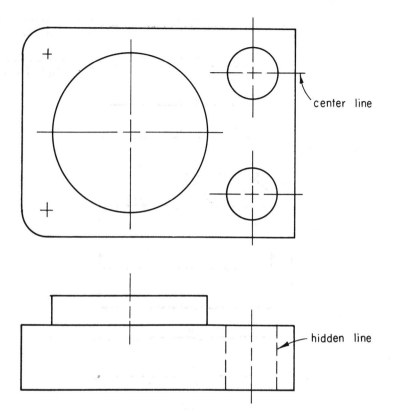

Figure 7–10 Use of hidden lines and center lines.

front views. The left front corner of this object has been rounded, and a + has been used to indicate the center of the round. Since this corner is less than half a circle, complete center lines are not used. Note that the center line consists of alternating long and short dashes, always beginning and ending with a long dash; and the short dashes always intersect at the center of the circle. A menu of various types of lines is shown in Figure 7–11.

Most engineering drawings consist of straight lines and circles. In pictorials, circles usually become ellipses, as shown in Figure 7–12. Samples of the basic shapes used in engineering drawing are shown in Figure 7–13. These basic shapes should be practiced in order to develop skill in all types of drawing.

The steps involved in sketching include both decision and drawing:

Step 1. Decide what object is to be sketched and what type of sketch is to be drawn. If a multiview sketch is chosen, how many views and which views will be drawn?

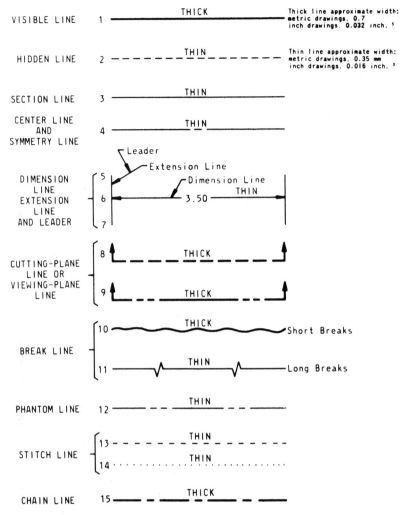

5   The metric line widths agree with ISO/DIS/128 (June 1977) and are not a soft metric conversion of the
    inch value.
    These approximate line widths are intended to differentiate between THICK and THIN lines and are not
    values for control of acceptance or rejection of the drawings.

Figure 7–11   Widths and types of lines. (ANSI 14.2 M–1979.)

Step 2. Decide what size and shape of paper is to be used for the sketch.

Step 3. Decide in what position the object will be placed (see Figure 7–6 for several examples) and which face of the object will be the front.

Step 4. Estimate the overall height, width, and depth of the object, and sketch these three dimensions on the paper. These dimensions form boxes which will exactly enclose the views.

Figure 7–12 Preliminary concept for heavy lift airship.
(*Courtesy* NASA.)

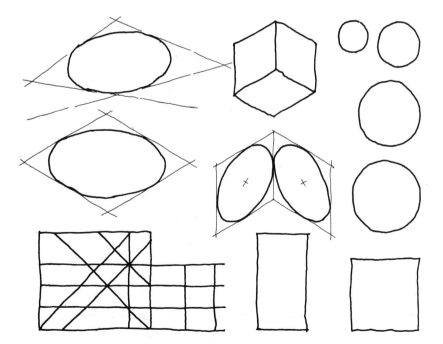

Figure 7–13 Common shapes in technical sketching.

Step 5. Identify specific features of the object and put them into all views. This step involves visualizing how each feature will appear from various directions, and carefully estimating proportions for the size and location of each feature.

Step 6. Darken all lines which belong in the three views, and if any light layout lines seem to detract, erase them. It is especially important to use dark lines if the sketch is on gridded paper, as it must stand out against the grid lines.

To implant these steps in your mind, an example follows which shows a specific application of the six steps.

EXAMPLE 7–1   Step 1. Decide to sketch three orthographic views of the object shown in Figure 7–14. The three views will be front, top, and right side.

Step 2. The paper size will be as shown in Figure 7–15a.

Step 3. The object will be oriented as shown in Figure 7–14, with surface B facing the front.

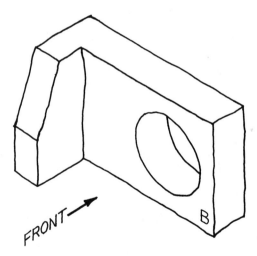

Figure 7–14 Three dimensional object for sketching exercise.

Step 4. In order to obtain the overall dimensions, lines are extended in Figure 7–15b. Points M and N are obtained in this manner.

The overall dimensions, height, width, and depth, are as shown in their proper placement in Figure 7–15c. Note that the height and width are only estimated once, but the depth has to be estimated twice. When locating the position of these boxes, the three horizontal spaces (x_1, x_2, x_3) should be approximately equal, and the three vertical spaces (y_1, y_2, y_3) should be approximately equal. Since this is a sketch and distances are estimated by eye (and not measured), spaces which are between one-half and double one another are generally considered acceptable. Poor spacing, such as that shown in Figure 7–15e, where x_3 is about six times x_1, should be corrected.

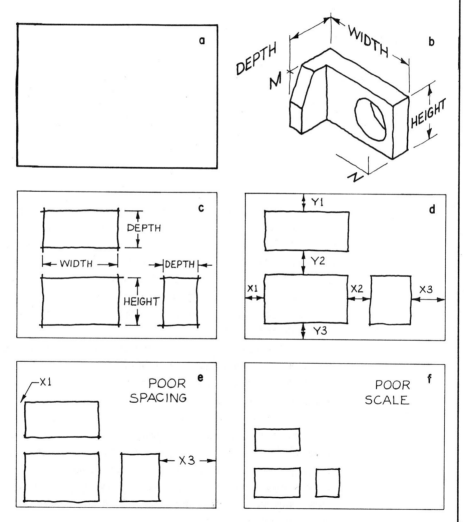

Figure 7–15 Steps for sketching orthographic views for a solid object.

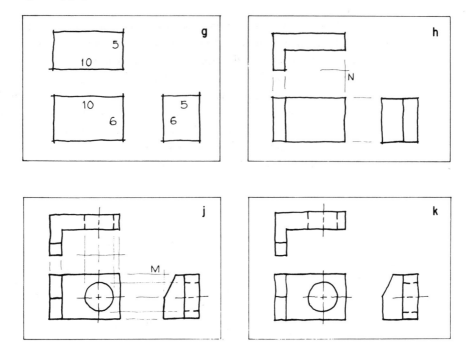

Estimating the size and the proportions of the object is probably the most important step in producing the sketch. If the object is drawn with too large a scale, the drawing will run off the edge of the paper. If the drawing is made too small, it looks lost, as shown in Figure 7–15f. Generally the size of the views should be approximately equal or slightly larger than the size of the spaces between views. Figure 7–15g shows a good choice of both scale and spacing.

Proportions are also very important. This particular object has a height, width, and depth in a ratio of approximately 6:10:5. Stated another way, the largest overall dimension is the width, the smallest is the depth. The width is about twice the depth, and the height is a little larger than the depth. If the sketch were being made on gridded paper, these proportions would be easy to set off. If the sketch is on plain paper, however, estimating the proportions takes practice, and several trial-and-error estimations may be made before a satisfactory layout is achieved. During this step, all lines are tentative and are therefore drawn extremely lightly.

Step 5. The object being sketched in Figure 7–15 has an "L" shape to it, with the right front corner (N) removed. The depth of the removed portion is slightly more than half of the total depth. The width of the removed portion is about 80% of the width. Putting this information into our sketch, we move from Figure 7–15g to Figure 7–15h. This feature shows most clearly in the top view.

A second feature of this object is the small angular cut which removes the upper front corner (M) from the object. This feature shows best in the right-side view and would be sketched in this view first. The cut extends down about half the overall height and is inclined at a slope of about 2:1, or approximately 60° with the horizontal. This feature is incorporated in Figure 7–15j. Note that the dimensions for each feature project from view to view.

A third feature to be included is the large circular hole. This hole is located about halfway up the large vertical surface set back parallel to the front of the object and equidistant from the bottom and the right side of the object. The diameter of the circle is about two-thirds the height of the object. This hole is indicated by a solid line (with appropriate center-line markings) in the front view and by dashed lines in the other views (see Figure 7–15j). Once all the specific features have been included in the sketch, all views should be checked for possible error.

Step 6. The final sketch of the object is shown in Figure 7–15k. The sketch has been checked, extraneous markings removed, and other lines darkened.

7.3 Orthographic projection systems

Systems of projection are used to create a two-dimensional image (such as a drawing) of a three-dimensional object. These systems are the geometry which explains how a particular type of drawing is formed.

In the projection system known as orthographic, (that is, projected perpendicular), the view or image of a three-dimensional object is created by enclosing the object inside a glass box with mutually perpendicular faces. Each of the three principal views (front, top, and side) is created by projecting all points on the object perpendicular to one of the three principal planes, as shown in Figure 7–16. Figure 7–17 shows the six views that are produced if all six principal planes are used. Rarely are all six views of an object necessary. Combinations of two or three of these views generally are sufficient to describe an object.

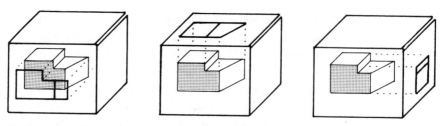

Figure 7-16 Projections of front, top, and right side views.

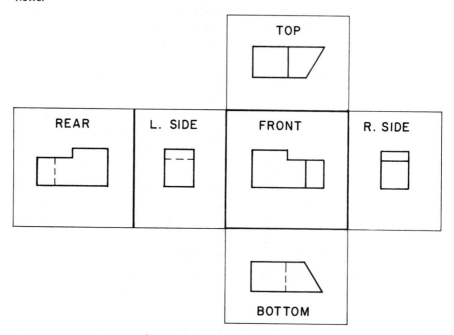

Figure 7-17 Conventional arrangement of all six principal views.

The views (or drawings) produced by orthographic projections are called multiview drawings, or orthographic views. In orthographic projection, all points are projected perpendicular to the projection planes, and the projection planes are all mutually perpendicular. If three-dimensional objects are to be described mathematically, a rectangular coordinate system is set up to allow the coordinates of each point to be defined. A frequently used set of axes is shown in Figure 7-18. The coordinates of point J are X_J, Y_J, Z_J. This set of axes has the relationship to the principal views shown in Figure 7-19: the top view is the XY plane, the front view is the XZ plane, and the right-side view is the YZ plane. Each view shows two coordinates. The measurement relationship is: $X =$ width, $Y =$ depth, $Z =$ height. Orthographic projections may also be drawn by a plotter controlled by a computer (see Figure 7-20).

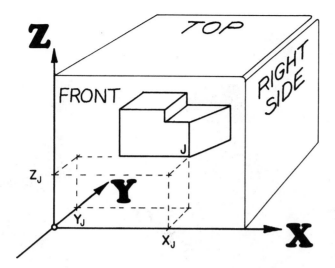

Figure 7–18 Three-dimensional rectilinear coordinate system.

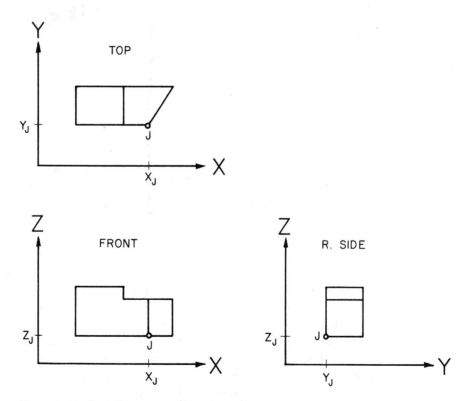

Figure 7–19 Coordinate system in orthographic views.

Figure 7–20   Plotting orthographic views. (*Courtesy* Chrysler.)

7.4   Pictorial views

The three types of pictorial views are *isometric, oblique,* and *perspective.* Each of the three is based on a different projection system, and each has its own particular advantages and disadvantages.

Isometric pictorials

Figures 7–6a and 7–14 are examples of isometric pictorials. These illustrations are produced by imagining the object placed so that all three principal faces make equal angles with the paper. Figure 7–21a, an isometric drawing of a cube, shows the top, front, and right side of the cube. If the hidden edges are included, as in Figure 7–21b, the hidden corner will appear exactly behind the front corner. The three mutually perpendicular edges of the object (OA, OB, and OC) appear at 90°, 30°, and 30° respectively with a horizontal line, as shown in Figure 7–21c. These three axes are the isometric axes.

Since the edges of the cube are not parallel to the paper, the length of the lines (e.g., OA, OB, and OC) in the isometric projection Figure 7–21c are less than their actual length (about 81.65% of the true length). In an isometric drawing, all three edges are usually drawn in their actual length. An isometric drawing (Figure 7–21d) is about 22.5% larger than an isometric projection (Figure 7–21c), and is easier to draw because all lines parallel to the isometric axes are drawn in their correct length.

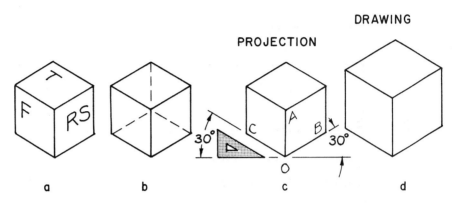

Figure 7–21 Isometric projection and isometric drawing.

EXAMPLE 7–2   An isometric drawing is to be made of the object shown in Figure 7–22. An appropriate procedure to follow would be:

Step 1. Decide which position to put the object in and what faces to show. Assume the object is in the position shown in Figure 7–22 and that the three faces which will be visible will be the top, front, and right side.

Step 2. Measure the overall height, width, and depth of the object, and set off these three dimensions (H,W,D) on the three isometric axes, as in Figure 7–23a. If the three dimensions were set off on the three axes shown in Figure 7–23b, the three faces would be top, front, and *left* side.

Step 3. Proceed from (a) to (c), constructing the enclosing box from which the final drawing will be "carved."

Step 4. Put individual features into the drawing. It is usually easiest to begin with features in the front and leave the rear features to

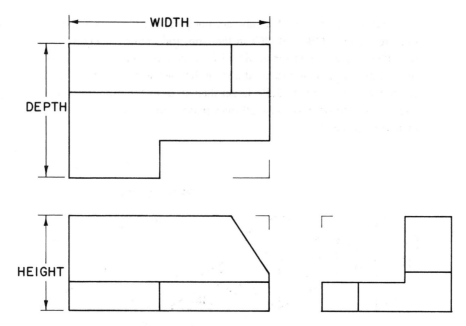

Figure 7-22 Object to be drawn in isometric.

last. The height and depth of the upper front portion which is to be removed are obtained from the right side view in Figure 7-22 for the removal of the upper front edge. These two distances, j and k, are measured in the isometric (see Figure 7-23d). Width m and depth n are determined from the top view in Figure 7-22, and the right front corner is removed, as shown in Figure 7-23e.

The last feature, the slanting surface, requires special attention. The two dimensions required to draw this feature, p and q, are determined from the front view in Figure 7-22 and are measured along a 30° and a vertical line, as shown in Figure 7-23f. It should be noted that the slanting line will not be the same length in the isometric drawing as it is in the multiview drawing.

This example points up some important rules of isometric drawing. First, only lines parallel to the three isometric axes can be measured. Lines not parallel to axes will appear either shorter or longer than those of the object being drawn. Second, angles do not appear in their true size in isometric drawings. An inspection of Figure 7-21 will reveal that all the 90° angles on this cube appear as either 60° or 120° in the drawing. Similarly,

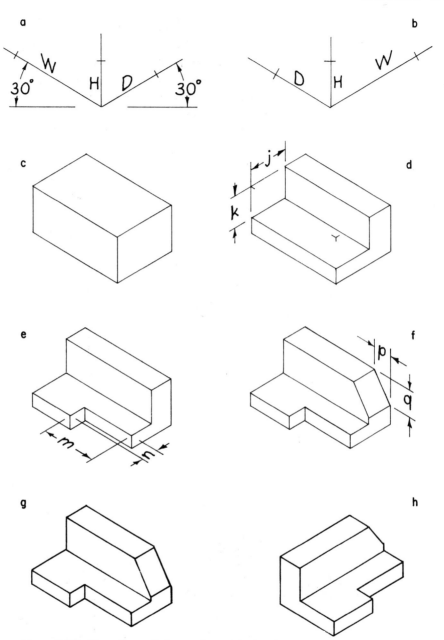

Figure 7–23 Steps in producing an isometric drawing.

the angular cut in Figure 7–23f does not have the same number of degrees in the isometric as it has in the multiview drawing, Figure 7–22.

Step 5. Darken the lines which form the view, and erase any construction lines if they are dark enough to interfere with the view. Figure 7–23g shows the finished view with all construction markings removed. Figure 7–23h shows the view which would result if the height, width, and depth were set off as in Figure 7–23b. This shows the left side of the object. The right-side view seems more descriptive for this particular object.

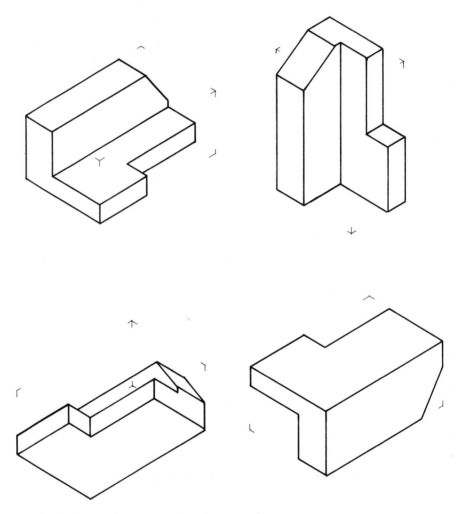

Figure 7–24 Four orientations of an object in isometric drawing.

As mentioned earlier, pictorial views include no hidden lines. If the object has a feature on the rear, bottom, or left side which does not show in this drawing, the object can be redrawn in a new position showing this feature. A very complicated object might require two or more pictorials in order to show all features.

To illustrate the effect of changing the object orientation, several other isometric views of the object from Figure 7–22 are shown in Figure 7–24. Note that some views give a better description than others.

Circular shapes and other curves occur frequently as part of objects. One additional example will show how a circular shape would appear in isometric.

EXAMPLE 7–3   Figure 7–25a shows two views of an object with a curve. In isometric drawings, a curve must be treated as a group of points. In this example, the curve will be treated as six points numbered 1 through 6. Each point will be located in the isometric drawing by finding its height, width, and depth from point T, the lower right front corner.

The initial steps are similar to those of the previous example. The enclosing box must be laid out as shown in Figure 7–23a–c, using the overall height H, width W, and depth D. To locate point 1 on the object in the isometric, distances X_1 and Y_1 are measured from point T on Figure 7–25a, and these distances are transferred to the box in Figure 7–26b. That is, point 1 is distance X_1 to the left of T, distance Y_1 above T, and zero distance behind T. The other points are similarly constructed (see Figure 7–26c).

The same curve exists on the rear face of this object, a distance D behind the front curve. Thus, points on the rear curve may be plotted by drawing a series of 30° lines through all six points on the front face and measuring off the distance D, the depth of the object. As shown in Figure 7–26d, some of the points may be hidden and are therefore omitted from the drawing. A straight line at 30°, tangent to both curves, should be included in the drawing (line RS in Figure 7–26d) as the back of the curve. A smooth curve should then be drawn through the six points on the front curve and the four points on the rear curve.

A circular hole, such as that shown in Figure 7–25b, may be located in the isometric drawing by imagining a square enclosing the circle, as shown in Figure 7–25c. The square is then located in the isometric, as in Figure 7–26e, and an ellipse fits inside the rhombus, as shown

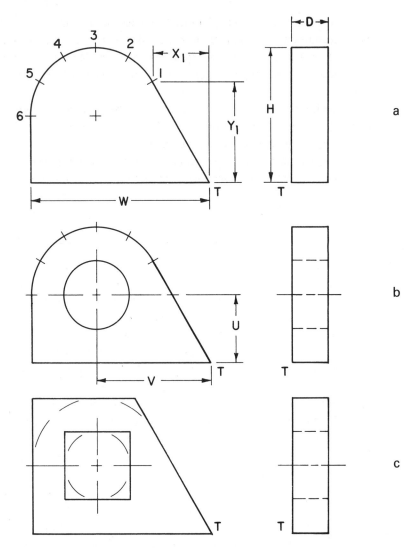

Figure 7–25 Curved object to be drawn in isometric and oblique.

in Figure 7–26f. Note that this ellipse is tangent to the midpoints of all four sides of the enclosing (rhombus) construction. Once the hole has been located in the front of the drawing, it is necessary to determine whether the exiting hole in the rear face will be partially visible. This is accomplished by putting the same size enclosing construction on the rear plane, which is a distance D measured up to the right on the 30° isometric line. The whole ellipse is shown in Figure 7–26g, but in the finished drawing, Figure 7–26h, only the visible portion of the rear ellipse

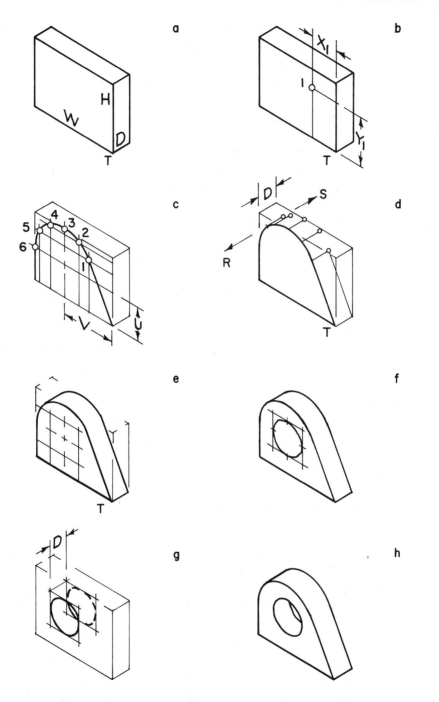

Figure 7–26 Eight steps in producing an isometric drawing.

is shown, and the hidden lines are omitted.

As will be seen in the next few pages, this object would be easier to draw using another type of pictorial.

Oblique pictorials

A second type of pictorial drawing is the oblique drawing. Oblique drawings are produced by imagining the object positioned such that one face is parallel to the paper, showing this face in its true shape. All other points on the object are projected onto the paper at any selected angle except 90° to the paper. This non-perpendicular, or oblique, projection gives the view its name. Several oblique drawings of a cube are shown in Figure 7–27. The front face is always drawn with vertical lines at 90° and horizontal lines at 0°. The receding axes, the third dimension (depth) which goes away from the viewer, can be drawn at any angle α upward to either the right or left. The scale on this receding axis can be the

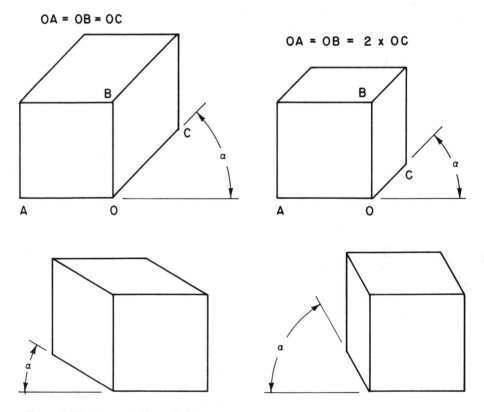

Figure 7–27 Four variations of oblique axes.

same as the scale on the other two axes, OC = OA = OB, or the scale on OC can be reduced. In Figure 7–27, the view at the top left shows the cube with a full scale on the receding axis. The view at the top right uses half scale on OC, that is, OC = ½(OB).

The steps for drawing an oblique pictorial are similar to those used for the isometric pictorial. The first step requires two decisions, however, which are not needed for an isometric drawing. The drawer must choose:

1. An angle for the receding axes, and

2. A scale for this axis.

EXAMPLE 7–4   Assume an oblique pictorial is to be drawn of the object shown in Figure 7–25b. The drawing will use an angle of 30° up to the right and will use full scale on the receding axis. Most of the steps involved in this oblique drawing are very similar to those used in the isometric.

Step 1. Decide in what position the object will be viewed and, particularly, which face will be parallel to the front. If circular shapes occur in a face, it is advantageous to use this face as the front of the object in the oblique drawing. This will allow the circles to be drawn with a compass and avoid the lengthy construction required for ellipses.

Step 2. Determine the overall height, width, and depth, and measure these distances on the three axes. This step is shown in Figure 7–28a.

Step 3. Begin putting individual features into the drawing. Measure two distances, V and U, to locate the center of the circular shaped curve in the upper left portion of the object. Use the actual radius and draw the circular shape with a compass, as shown in Figure 7–28b. Repeat this same circular shape on the rear surface, locating the rear center by measuring depth D along the receding axis through the front center (see Figure 7–28c). Construct lines tangent to these two circular arcs and extending to point T and its rear counterpart.

Next, draw the circular hole on both the front and rear faces, and note whether or not the rear circle will be visible (Figure 7–28d).

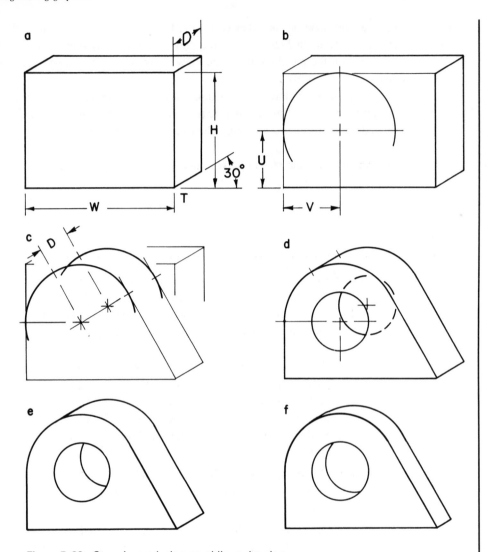

Figure 7–28 Steps in producing an oblique drawing.

Step 4. Darken visible lines and eliminate any lines not actually part of the view. The resultant drawing is shown in Figure 7–28e. For comparison, the same object is shown in Figure 7–28f with the scale on the receding axis one-half that of the full scale used in Figure 7–28e.

Perspective pictorials

The third type of pictorial drawing is perspective. For many years, artists struggled to develop a system to represent reality in their artwork. Perspective drawing is the result of their efforts and is the most realistic pictorial drawing. A perspective could be constructed by drawing what is observed through a pane of glass onto the glass itself. A major feature of perspective is the concept that parallel lines which recede from the viewer appear to converge. Stated another way, the size of objects will become smaller as the objects are placed further from the viewer.

Again using the familiar cube as an example, three perspective sketches are shown in Figure 7–29. The first example on the left (a) shows a cube with one face parallel to the paper on which the image is being projected. The second example (b) shows the cube if only one edge, OB, is parallel to the paper. Both views (a) and (b) show the cube with the top surface horizontal, and thus the vanishing points U and V lie on the horizon. The third example (c) shows a cube tilted so the top is no longer horizontal, and none of the three vanishing points W lies on the horizon.

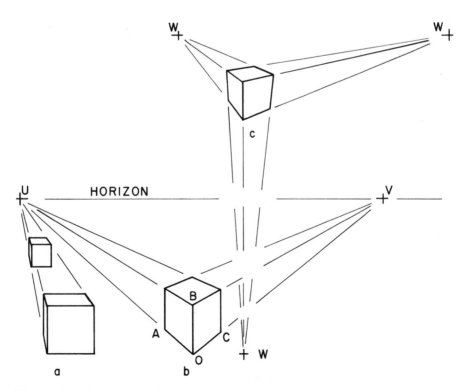

Figure 7–29 One, two, and three point perspective.

These three examples are known as one-, two-, and three-point perspectives, respectively. In engineering drawing, one- and two-point are the most common. Three-point perspectives are used only in cases where an object with a great height is being depicted.

All three images in Figure 7–29 may be explained as being projected onto the paper from a single point, the observer's position, as shown in Figure 7–30. If the object is in front of the paper, the drawing is larger than the object, as in object A. If the object is placed behind the paper, the image (drawing) is smaller than the object, as in object B. The further the object is from the observer, the smaller its image will be.

Perspective drawing requires a good deal of preliminary construction, mainly to establish correct lengths. Since the sizes change with depth, no constant scale can be used along any axis. Because of the lengthy construction required for perspective drawing, this text will consider only perspective sketching.

One simple approach to perspective sketching is to incorporate some perspective features into the other two types of pictorials, isometric and oblique. For relatively small objects such as machine parts, such a procedure is quite satisfactory. For large objects such as buildings, more detailed construction is necessary.

View (a) of Figure 7–31 shows the isometric drawing constructed previously, and view (b) shows the same drawing sketched with a perspective "flavor," with a pseudo two-point perspective. The same object is shown in an

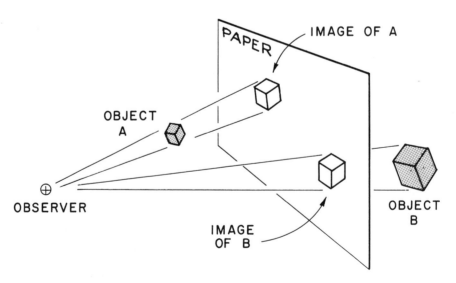

Figure 7–30 Perspective projection.

oblique drawing in view (c). View (d) shows the oblique drawing changed to a one-point perspective sketch.

One important consideration in perspective is the height of the observer relative to the height of the object being viewed. For all of the perspective sketches so far (Figures 7–29 and 7–31) the viewer has been above the object, looking down, and therefore the top of the object has been visible. If the object is taller than the viewer, the top surface will not be visible. Figure 7–32 shows three objects with the following heights: (a) about half the height of the viewer, (b) about the same height, and (c) about twice the height.

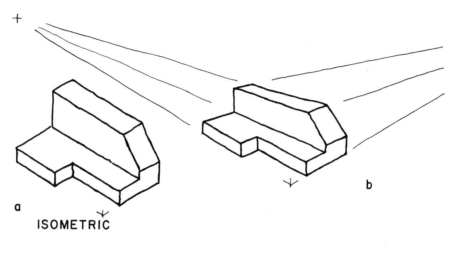

ISOMETRIC

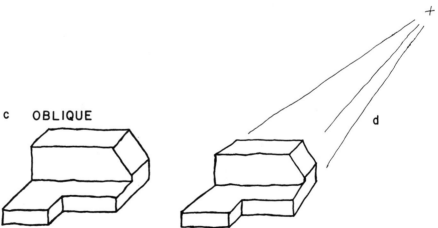

c   OBLIQUE

Figure 7–31 Pseudo perspective in isometric and oblique sketches.

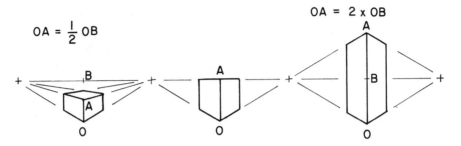

Figure 7–32 Height estimation in perspective projection.

The scale on the receding axis must also be considered in perspective sketches. For example, the distances between fence posts must be estimated for the perspective in Figure 7–33. For all the posts parallel to the horizon, the space between posts, X, is approximately equal to the height of the posts. When the fence turns the corner and recedes away from the observer toward the horizon, the spaces become smaller just as the post heights become smaller. Thus, about halfway to the horizon when the posts are about half as tall as the front posts, the space, Y, is approximately half of the front space, X.

Circular shapes in perspective are generally ellipses and should be boxed in a manner similar to that shown in Figure 7–26e and 7–26f. The size of this en-

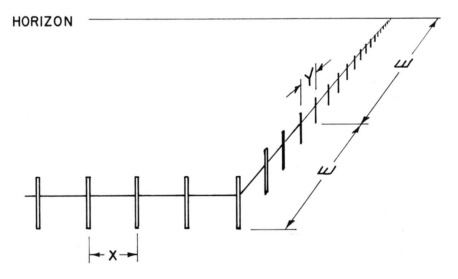

Figure 7–33 Perspective heights and the horizon.

closing box must be estimated by eye in perspective, however, because the scale is not easily measurable. Note that in Figure 7–34a the enclosing box is not a rhombus, but just a four-sided polygon. The completed perspective sketch without construction is shown in view (b). If the circle is parallel to the front of an object in a one-point perspective, however, the circle will appear circular, rather than elliptical, in the perspective drawing. Such a circle can be drawn easily with a compass, as with the hollow cylinder shown in Figure 7–34c.

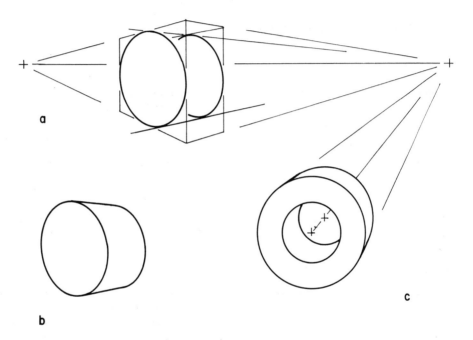

Figure 7–34 Circular shapes in perspective.

The three types of pictorial drawings are summarized and compared in Table 7–2. The most realistic drawing is the most difficult to construct—the perspective (see Figure 7–35). In sketching, however, some perspective realism can easily be incorporated into sketches without any extensive construction.

Table 7–2 Pictorial Drawings

Feature	Isometric	Oblique	Perspective
3 Axes	vertical, 30° with horizontal; no exceptions	horizontal, vertical, and choice of angle for receding axis	no specific axes
Scale	full scale on all axes	full scale on horizontal, vertical; choice of scale on receding axis	scale varies with distance
Parallel lines on object	parallel in view	parallel in view	usually converge; exception in one-point perspective in front plane
Circles on object	ellipses	ellipses, unless parallel to front face	ellipses, unless parallel to front face in one-point perspective
Vertical lines on object	vertical	vertical	vertical except in three-point perspective
Realism		poorest	best
Work required to construct		least work	most work
Hidden lines	omit	omit	omit
Angles	all distorted	shown true if parallel to front; otherwise distorted	shown true if parallel to front in one point perspective, otherwise distorted

If pictorial drawings are desired as computer output, the images are obtainable by the following methods. First, a coordinate system must be established in order to allow the description of the object to be input into the computer. Let us assume that the object is described in terms of the X, Y, and Z coordinates of all points, with the axes as shown in Figure 7–36.

Figure 7–35 Perspective view showing parallel lines converging in a subway station. (*Courtesy* CRS Group, Inc.)

If an isometric projection (slightly smaller than an isometric drawing) is desired, the object should first be rotated 45° about the Z axis, and then rotated 35°16′ about the X axis. After these two rotations, the Y coordinate is ignored, as depth does not change the size of the image in isometric projection. The image to be plotted, then, corresponds to the X and Z coordinates of each point.

To obtain an oblique drawing, the X and Y coordinates of each point are translated in a predetermined direction (the direction of the receding axis) a distance proportional to the Z coordinate (the depth of the point). The new X and Y coordinates are then plotted.

To obtain a perspective view, an object with no rotation will produce a one-point perspective. After one rotation, a two-point perspective will result, and after two successive rotations about two different axes, a three-point perspective will be produced. Information as to the distance the observer is from the object and the plane on which the image will be projected must be furnished to the computer. If the paper is assumed to be the XY plane, as shown in Figure 7–36, and if the observer is stationed a distance D from this plane along the Z axis, then the image

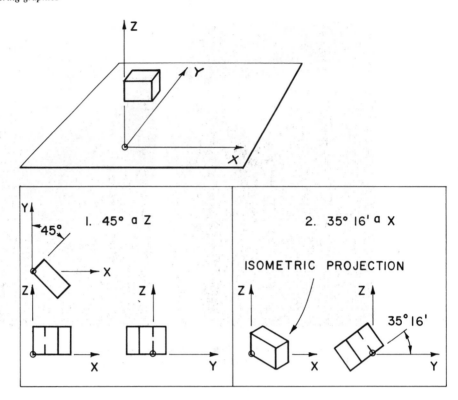

Figure 7–36 Obtaining an isometric projection in computer graphics.

of the object is obtained by multiplying the X and Y coordinates of each point by a factor proportional to the depth of the point, the Z coordinate. This projection is the only pictorial in which the third coordinate affects the scale or size of the image.

7.5 Conventions, standards, and practice

Many rules for engineering drawings have become established through common practice over long periods of time and ultimately become formalized in drawing standards.

Sectioning

Sectioning is a widely used convention which involves an imaginary cut into an object to allow internal features to be shown more clearly. This convention is almost as old as drawing itself and was used by Egyptians in the construction drawings for the pyramids. Its use is widespread, and probably over half of all technical drawings use some form of sectioning.

To illustrate some of the many types of section drawings used in engineering, a simple hollow cylinder will be used. In Figure 7–37a the cylinder is shown

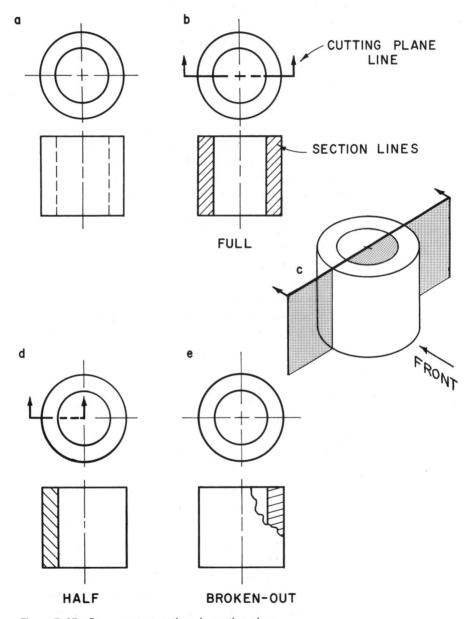

Figure 7–37 Common conventions in section views.

unsectioned. The same cylinder is shown as a full section in view (b). This drawing demonstrates the use of two different types of lines: (1) a cutting plane line, with arrowheads on the ends, which indicates the location of the imaginary cut, and (2) the section lines, or crosshatching, which show in the sectioned front view where material exists (and would be cut). The cutting plane line is a thick black line with alternating one long and two short dashes. The arrows on the ends of the cutting plane indicate the direction in which the observer is looking at the cut section. When a symmetrical object is cut along a plane of symmetry, as in this case, the cutting plane line is sometimes omitted. The section lines are thin black lines, parallel and equally spaced. They may be drawn at any slanting angle, but not horizontally or vertically. For a single object, all section lines should be drawn at the same angle.

Figure 7–37c shows the imaginary cutting plane for this full section. Figure 7–37d shows the same object as a half section. A half section is used only with symmetrical objects and cuts only halfway through the object. The left half of the front view is imagined cut in section, while the right half is not. Note that *no* hidden line appears in the right half. Generally, no hidden lines are used in sectioned views. The boundary between the sectioned and the unsectioned half is a center line, though a solid line is sometimes used.

A third type of sectioned view is shown in view (e). This section is called a broken out, and the jagged line which serves as a boundary for the sectioned portion is a "break" line. This indicates an imaginary break in the outer wall, allowing the viewer to see the interior. No cutting plane line is drawn here as the object is symmetrical, and it is assumed to be cut through the plane of symmetry. However, if any doubt exists as to where the imaginary cut was located, a cutting plane could be drawn in a manner similar to the one shown in the top view in Figure 7–37d.

A longer cylinder is used to demonstrate other section views. Figure 7–38 shows a revolved section which is produced by imagining a plane cutting through the object at the location of the center line and then revolving it 90° so the cross-sectional shape at this point can be seen. This section view is drawn on top of the existing view, and no cutting plane line is used.

Figures 7–38b and 7–38c are variations of revolved sections in which the sectioned view is not drawn on the existing view, but instead the view is "broken" to allow the sectioned view to be drawn on clean paper, not congested by the existing view. The break line used in Figure 7–38b is a general purpose break line. The break used in Figure 7–38c is a break specifically used for cylindrical objects and is drawn in the shape of a figure 8 with one-fourth of the 8 missing.

A removed section, such as that shown in Figure 7–38d is produced by a cutting plane perpendicular to the cylinder's axis and then rotated and removed to a location alongside the existing view. This type of section is identical to a re-

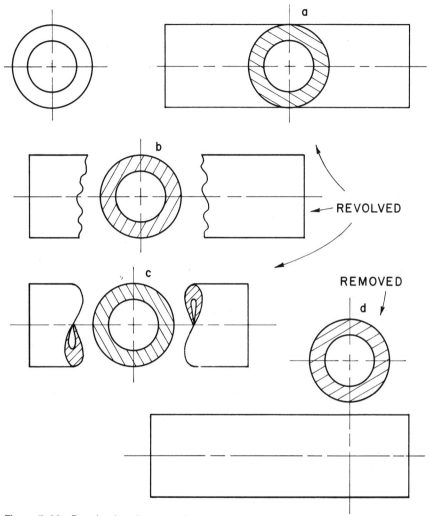

Figure 7-38 Revolved and removed sections.

volved section, except that the section itself is moved off the view.

Figure 7–39 shows a variety of breaks which are used on engineering drawings.

A typical example using removed sections is shown in Figure 7–40. Clearly, this type of section is very useful in depicting how an object changes shape along an axis. Without sectioning, a description of this type of object would be extremely difficult.

Figures 7–41 and 7–42 illustrate some other variations of sectioning. The top view in Figure 7–41 is an offset section of the object shown in the lower

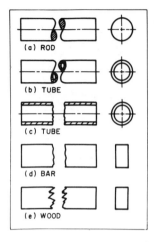

Figure 7–39 Conventional representation on breaks in elongated parts. (ANSI Y 14.3–1975.)

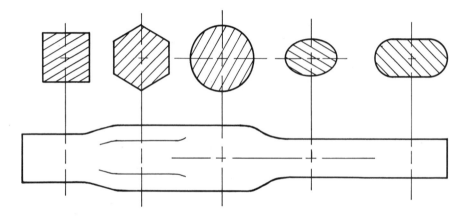

Figure 7–40 Removed sections.

views. An offset section is similar to a full section, but the cutting plane is offset about midway through the object to include a feature it would miss if it continued straight through. Note that the offset in the cutting plane does *not* produce a line in the sectioned top view.

An aligned section view is shown in Figure 7–42. The aligned section is produced by a cutting plane which has a bend in it but is drawn "straightened out." It is drawn straight to show the full length and to avoid congestion.

Many complex objects will require several sections to give the observer a complete shape description. In such cases, the various cutting planes are usually identified with letters, as shown in Figure 7–43. It should be noted that when a relatively thin wall of material is cut by a cutting plane, this material is not sec-

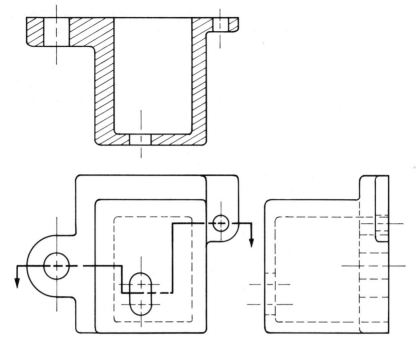

Figure 7–41   Offset section.

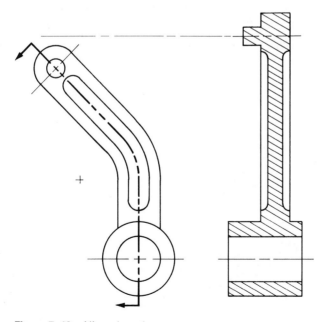

Figure 7–42   Aligned sections.

tioned. In Figure 7–43, cutting plane *A-A* cuts through an interior vertical wall. Since this thin material is imagined cut even thinner by cutting plane *A-A*, it does not receive crosshatching in the section view *A-A*.

Sectioning is also used in pictorial views such as isometric, oblique, and perspective. Figure 7–44 shows an isometric section of the object shown in Figure 7–43. This is a partial section, since only part of the object is sectioned. In a

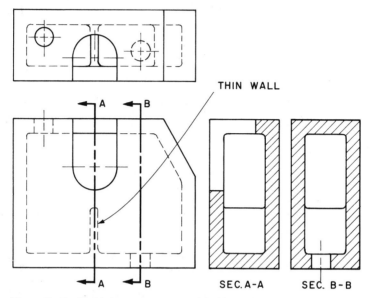

Figure 7–43 Multiple sections, one with thin wall.

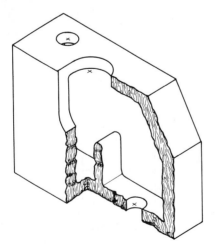

Figure 7–44 Pictorial section.

partial section, some portion of the object, large or small, is imagined to be cut away. Figure 7–45 shows a broken-out section of a complex object, the spacelab in the payload bay of the space shuttle.

Figure 7–45   The space shuttle. (*Courtesy* NASA.)

All the illustrations so far in this chapter have involved a single object. In engineering, however, frequent use is made of assembly drawings in which several parts are shown as they normally fit together. Sectioning is often used in such assembly drawings. When more than one part is shown in section, each particular part should be crosshatched with section lines appropriate for the material from which the part is made.

Figure 7–46 shows the symbols commonly used for sectioning materials. An example of the use of these markings in an assembly drawing is given in Figure 7–47, and Figure 7–48 shows the preparation of an assembly drawing using a drafting machine.

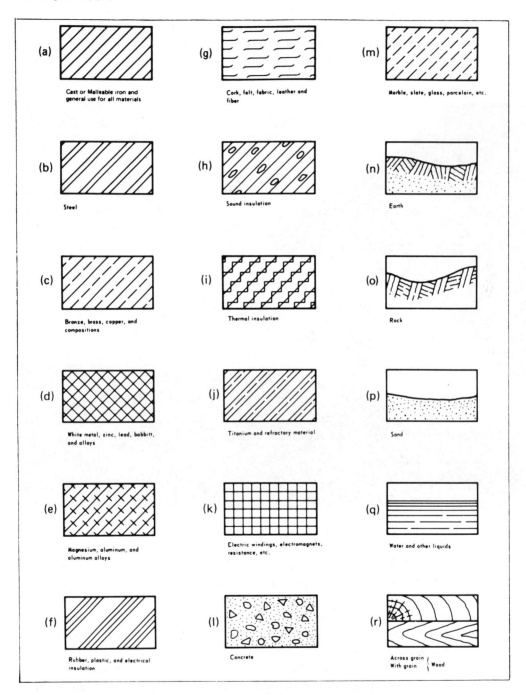

Figure 7–46 Conventional representation of materials in section. (ANSI Y 14.2M–1979.)

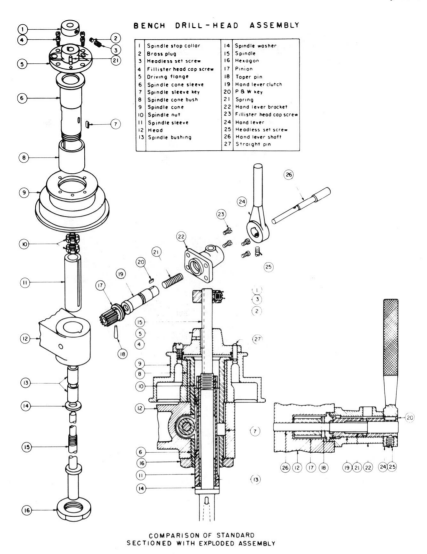

BENCH DRILL-HEAD ASSEMBLY

1	Spindle stop collar	14	Spindle washer
2	Brass plug	15	Spindle
3	Headless set screw	16	Hexagon
4	Fillister head cap screw	17	Pinion
5	Driving flange	18	Taper pin
6	Spindle cone sleeve	19	Hand lever clutch
7	Spindle sleeve key	20	P & W key
8	Spindle cone bush	21	Spring
9	Spindle cone	22	Hand lever bracket
10	Spindle nut	23	Fillister head cap screw
11	Spindle sleeve	24	Hand lever
12	Head	25	Headless set screw
13	Spindle bushing	26	Hand lever shaft
		27	Straight pin

COMPARISON OF STANDARD
SECTIONED WITH EXPLODED ASSEMBLY

Figure 7–47 Assembly drawings: exploded assembly and two sectioned subassemblies. (ANSI Y 14.4–1957.)

7.6 Auxiliary views

Auxiliary views are used in multiview drawings to provide a more complete description of an object. Many three-dimensional objects have peculiar shapes which are described more easily and more clearly in auxiliary views. An example

Figure 7–48 Preparing an assembly drawing. (*Courtesy* K & E.)

of an object of this nature is shown in Figure 7–49 in pictorial and in Figure 7–50 as a multiview drawing. The two lines, labelled *HF* and *FP*, which are placed between the views in Figure 7–50 are called reference lines. Each reference line represents a 90° fold in the paper, and each reference plays a very important part in understanding and constructing auxiliary views.

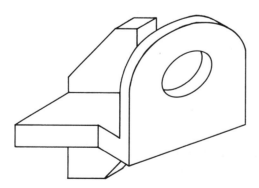

Figure 7–49 Typical object requiring auxiliary view.

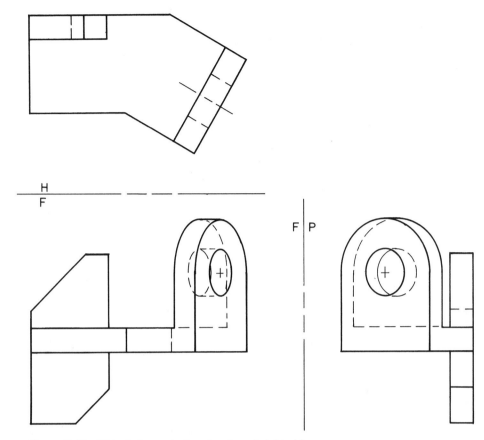

Figure 7–50 Object shown in top, front, and right side views.

An auxiliary view would be useful in describing the object shown in Figure 7–50 because both the front view and the right side view have several visible and hidden ellipses. These ellipses are tedious to draw and do not show the true shape of the vertical semicircular feature on the right end of the object. Figure 7–51 shows the top, front, and right-side views, and an auxiliary view. This auxiliary view is a view looking directly through the circular hole in a direction perpendicular to the angled surface on the right end of the object. The auxiliary view is projected onto a plane perpendicular to the top view. The relationship between the front, top, side, and auxiliary planes is shown in Figure 7–52. Note that in all cases, projection lines are perpendicular to the reference lines, and that height X is common to both the front view and the auxiliary view.

Further inspection of the three views in Figure 7–51 reveals that they contain some unnecessary detail. Since the right end of the object does not show clearly in the front view, this portion could be omitted to save drawing time and

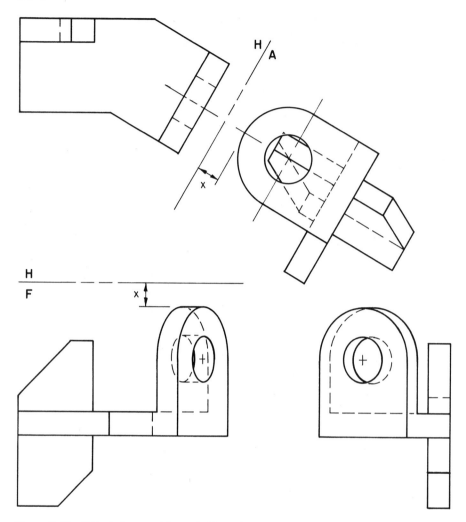

Figure 7–51 Object shown with an auxiliary view added.

to eliminate confusion and possible error. Some detail in the auxiliary view also is distorted and could be omitted. The simplest, clearest, and therefore the best drawing of this particular object would be the version in Figure 7–53. This version shows a complete top view, a partial front view, and a partial auxiliary view. Compare this to Figures 7–50 and 7–51.

The construction of auxiliary views involves the same sequence of steps necessary to construct a third principal view, changing a two-view drawing to a three-view drawing. To illustrate the similarity of these steps, a right-side view and an auxiliary view will be added to the object shown in Figures 7–54a and 7–54b. Although this relatively simple object would not ordinarily require an auxiliary view, its simplicity will make visualization of the process easier.

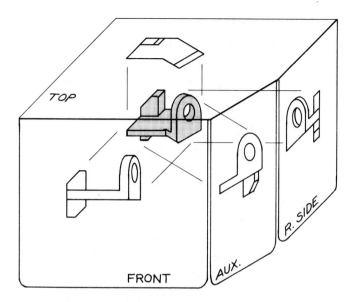

Figure 7–52 Projection which produces an auxiliary view.

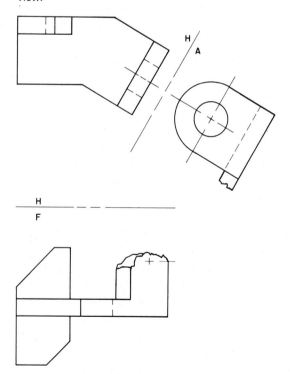

Figure 7–53 Top, partial front, and partial auxiliary view.

EXAMPLE 7–5   The procedure for drawing an auxiliary view:

Step 1. Decide which new view is to be drawn. Before any additional views are drawn, at least two complete views must be known. In Figure 7–54b the two given views are a top and front view. The two new views to be drawn are a right-side view and an auxiliary view taken so the viewer will be looking in a direction normal (perpendicular) to the inclined surface *G*. Both new views will be projected off of the front view in this example. The two arrows in Figure 7–54c show the viewing directions which have been established.

Step 2. Draw reference lines perpendicular to the lines of sight (arrows). A reference line should be drawn between the two given views if it is not already in the drawing.

Step 3. Project every point on the object into both new views, as shown in Figure 7–54d.

Step 4. Locate each point in the new view by measuring its distance from the reference line. This distance is obtained in the view next to the view from which the points were projected. In this example, the distances are measured in the top view since the top view is next to the front view, and the points were projected from the front view.

In order to keep track of the points, it is necessary to identify a specific point in the top and front view, and then set off this point's distance on its projection ray into each new view. It is important to be able to visualize the three-dimensional object, and it may also be helpful to number the points involved.

The pictorials in Figure 7–54a show the eight points numbered, with 1, 2, 3, and 4 on the front face and 5, 6, 7, and 8 on the rear face. These points have been added to the drawing in Figure 7–54e. In Figure 7–54f, point 1 has been located in the auxiliary view and the right-side view by obtaining distance *X* in the top view and setting off this same distance (measured from the reference line) in both new views. Figure 7–54g shows all eight points transferred into both new views. It should be noted that if the auxiliary view can be visualized and drawn without resorting to the use of numbers, or with numbering some of the points, the drawing can be more quickly and easily completed. The numbering of points may be omitted as visualizing capabilities increase.

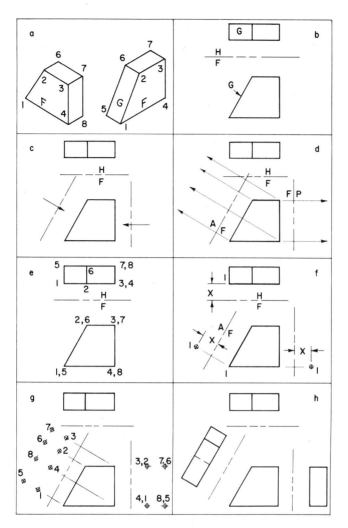

Figure 7–54 Eight steps in constructing an auxiliary view.

Step 5. Finally, connect the eight points with lines. Again, it is important to be able to visualize the object being drawn. Some lines may be visible because of the angle from which the object is being drawn; others may be hidden. The only hidden line in this example is the line connecting points 4 and 8 in the auxiliary view. Since the 4–8 edge cannot be seen when the front view is observed from the direction of the arrow, this line is drawn as a hidden line. Figure 7–54h shows the completed drawing.

Auxiliary views are often used to obtain the true shape of a curve on a three-dimensional object. Figure 7–55a shows a circular duct intersecting two vertical planes, M and N. The hole in plane N is a circular arc, and its true shape is shown in the front view. The hole which is cut in plane M appears circular in the front view, but its true shape is an ellipse and can be obtained by drawing an auxiliary view.

In Figure 7–54 the auxiliary view normal to plane G yielded the true shape of that surface. In Figure 7–55 an auxiliary view normal to plane M will show

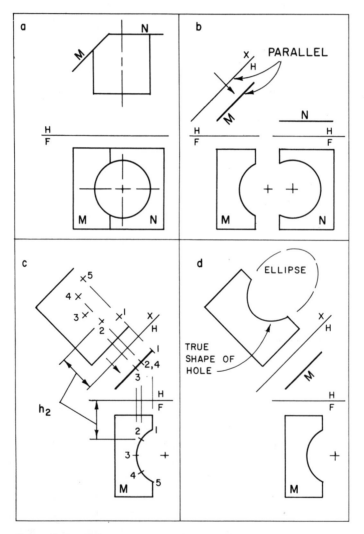

Figure 7–55 Four steps in constructing a curve in an auxiliary view.

this plane's true shape. This auxiliary view is drawn with the direction of sight perpendicular to plane M, as shown in Figure 7–55.

The major difference between this and the previous example is the necessity of dividing the curve in the front view into a series of points. Since this is a relatively smooth curve, only five points were used: the end points 1 and 5, and three intermediate points. If the curve were larger or more complicated, more points would be used.

After the five points are chosen in the front view, they are projected to the top view and from there to the auxiliary view. Points are always projected perpendicular to the reference lines. To locate the points on their projection rays in the auxiliary view, each point's distance from its reference line is measured in the front view and transferred to the auxiliary view. For example, in Figure 7–55c height *h* is measured from the reference line down to point 2 in the front view; this distance is then set off in the auxiliary view, measuring from the reference line along the projection ray from point 2. After all five points are located, a smooth curve is drawn through them, as shown in Figure 7–55d, forming part of an ellipse. This view could be used as a pattern for cutting the hole for the circular cylinder to enter.

Auxiliary views are useful not only to show the true shape of a plane, but also to find the true length of a line.

Figure 7–56a shows three views of a truncated pyramid. In order to determine which lines are of true length in a specific view, it is necessary to apply the following rule: a line is true length in a view if this line is parallel to a reference line (or appears as a point) in an adjacent view. Applying this rule to Figure 7–56a, *XY* is true length in the front view because *XY* is parallel to the reference line in the top view. Similarly, *YZ* is true length in the auxiliary view and in the top view because it appears as a point in the front view. Table 7–3 shows which lines are true length in which views. Study the table and its results.

The auxiliary view shows the true lengths of *DZ* and *CY,* lengths which are not available in the principal views. Line *WA* does not appear true length in any of the views. To obtain the true length of *WA,* a reference line would be drawn parallel to *WA.* The auxiliary view drawn using the new reference line would show the true length of *WA.* Since *WA* appears in both the top and front views, two such reference lines are possible [lines *H*3 and *F*2 in view (c)]. The completed drawings in view (d) yield the true length of *WA.* The results of the two, of course, are identical.

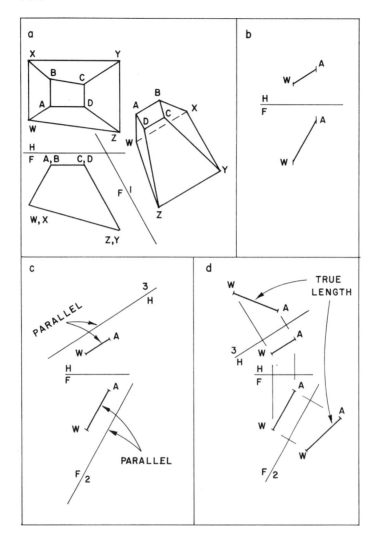

Figure 7–56   Finding the true length of a line in space.

Auxiliary views are used in drawings to obtain views from specific vantage points or directions of sight. This approach generally assumes that the object remains stationary and the viewer moves to a new position. The same results may be achieved by assuming that the viewer remains stationary and the object rotates to a new position. In computer graphics, views usually are obtained by specifying how the object is to be moved. Typically, the object is rotated about the X, Y, or Z axis, or about a combination of these. (See Figures 7–18 and 7–19 for standard positioning of these axes.)

Table 7–3 Identifying True Length Lines*

Line	True Length in		
	H (top)	F (front)	I (auxiliary)
AB	X		X
AD	X	X	
BC	X		
DC	X		X
AW			
BX			
CY			X
DZ			X
WX	X		X
XY		X	
YZ	X		X
ZW			

*True lengths are checked (X).

Figure 7–57 compares these two methods of describing an object. A front and right-side view of an angular object is shown in view (a). The auxiliary view approach would result in the partial auxiliary shown in the upper drawing. The rotational approach would result in the rotation of the object 30° counterclockwise (in the right-side view) about the X axis. The true shape of the inclined surface is evident in the front view, not in an auxiliary view, as shown in the lower drawing.

The rotational method used in computer graphics does not use an auxiliary view. Instead, two front views are used, one drawn before rotation, and one drawn after rotation. Both front views are obtained by plotting the X and Z coordinates of each point on the object. When the object is rotated, these coordinates change and a new view is created, usually by matrix multiplication of data stored in the computer. The traditional auxiliary view approach uses one front view and an auxiliary view to supply the same information.

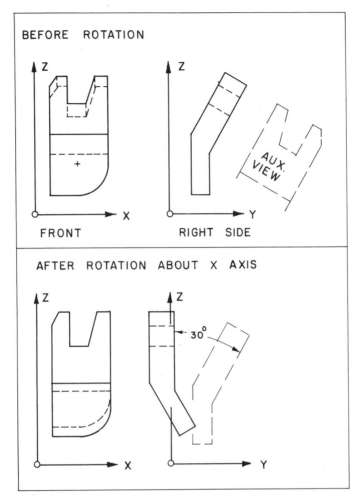

BEFORE ROTATION

Z

FRONT

Z

RIGHT SIDE

AUX. VIEW

AFTER ROTATION ABOUT X AXIS

Z

X

Z

30°

Y

Figure 7–57 Rotating an object about the *X* axis.

7.7 Assembly drawings

As mentioned earlier, assembly drawings are often used in engineering to show the manner in which parts normally fit together. Figure 7–47 shows two examples of assembly drawings. The example on the left is an exploded assembly, showing each part displayed in three-dimensional space in the order in which the parts are assembled. This particular exploded assembly is an isometric drawing. The example on the right shows all the parts assembled, using a sectioned view to show interior details. Note that part 12 in this drawing is shown twice, once in the main sectioned view and a second time to the right, showing the handle subassembly. A complicated assembly, such as that of an automobile, may be broken down into subassemblies (e.g., the rear wheel assembly) and subassemblies of subassemblies (e.g., the brake mechanism for the rear wheel).

The purpose of most assembly drawings is to show how a set of parts fit together or how a particular device works. They are a useful guide to assemble or disassemble a device. They do not give the details necessary for manufacturing individual parts and usually do not include dimensions. Assembly drawings do have identifying numbers so individual parts may be located in the device. Often a parts list is included giving the number, name, and perhaps some additional information concerning each part. For example, in Figure 7–47, four cap screws, part number 23, are needed for each bench drill-head.

Assembly drawings are used in many different ways. Elaborate, multicolor assembly drawings are sometimes used in training personnel to operate and/or repair a particular piece of equipment. Designers often use drawings similar to assembly drawings as vehicles for idea integration. Specific sizes and shapes of individual parts may be determined by means of such drawings. Once the individual parts have been designed and drawn in separate multiview drawings, known as detail drawings, another assembly drawing may be made using the information on the detail drawings in order to ensure that all the various parts fit together correctly. Assembly drawings frequently are furnished to purchasers of equipment to enable them to order new parts if repair becomes necessary.

A patent drawing is a special type of assembly drawing intended to describe clearly the ideas incorporated in the patent. An example of a patent drawing is shown in Figure 7–58.

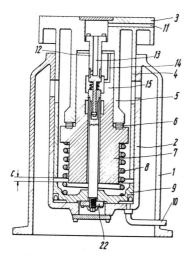

Figure 7–58 Patent drawing. (*Courtesy* U.S. Patent Office.)

7.8 Scales

Most drawings are made to a specific scale. If the drawing is exactly the same size as the object being drawn, the drawing is called a full-scale drawing. If the drawing is made only half the size of the object, the drawing is referred to as half-size or a half-scale drawing. Freehand sketches usually are not drawn to a specific scale; instead, approximate scales are used as proportions are estimated.

The United States currently is in the process of changing to the metric system, and in the future, most drawings will be made using metric scales. The two scales presently used on engineering drawings in the United States are the metric scale and the Engineer's scale, which uses decimal inches. These two systems are compared in Figure 7–59 and Table 7–4.

A triangular Engineer's scale usually has six scales, labelled 10, 20, 30, 40, 50, and 60. These numbers indicate how many divisions per inch each scale contains.

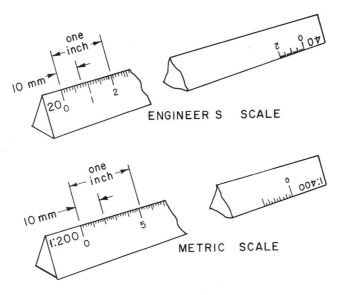

Figure 7–59 Comparison of an Engineer's scale and a metric scale.

EXAMPLE 7–6   If a map were being drawn to a scale of one inch (on the paper) representing 300 miles, the 30 scale would be used. Each inch would represent 300 miles, and since this scale has 30 divisions per inch, each division would represent 10 miles. Figure 7–60 shows an example in which a distance of 640 miles is to be measured using the scale $1'' = 300$ miles. The result is evident from the drawing. If the distance to be measured had been 645 miles, the additional five miles would have to be estimated by eye, as the smallest division on this scale is 10 miles.

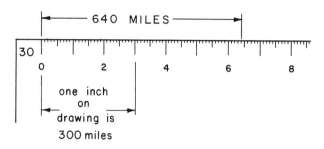

Figure 7–60 Using the Engineer's scale.

Table 7–4 Scale Comparison

| Scale | Engineer's | | | Metric | | |
| Name | Label | Smallest Division | | Label | Smallest Division | |
		inches	mm		inches	mm
Full	10	0.100	2.54	1:100	0.039	1.00
Half	20	0.050	1.27	1:200	0.039	1.00
Third	30	0.033	0.85	1:300	0.026	0.67
Quarter	40	0.025	0.63	1:400	0.049	1.25
Fifth	50	0.020	0.51	1:500	0.039	1.00
Sixth	60	0.017	0.42	1:600	0.033	0.83

Table 7–5 demonstrates the application of the Engineer's scale to a variety of problems. Each scale is expressed as $1'' = x$, where x can be any number with a leading digit of 1, 2, 3, 4, 5, or 6. This leading digit corresponds to the leading digit of the scale selected. For example, if $x = 5000$, the leading digit is 5 so the 50 scale is selected. The user must furnish the decimal points and appropriate zeros, keeping in mind that one inch here represents not 50 but 5000 miles on the drawing.

Triangular metric scales come with a variety of different scale divisions. Table 7–4 lists six common scales for the triangular metric scale as well as the Engineer's scale. The metric scales shown have the same leading digits as the six Engineer's scales. It should be noted that some metric scales have a 1:250 in place of the 1:600 scale. All scales expressed on metric drawings are actually dimensionless ratios of the size of the drawing to the size of the object. Thus a 1:200

Table 7–5   Engineer's Scale Applications

Scale of Drawing	Scale Selected	Typical Application	Smallest Division in Scale
1″ = 10 ft	10	a property map	1 ft
1″ = 5000 mi	50	a world map	100 mi
1″ = .5°C	50	an axis on a graph	.01°C
1″ = 30000″	30	a city map	1000″
1″ = 1″	10	a machine part, full scale	.1″
1″ = 2″	20	a machine part, half scale	.1″

scale means that 1 mm on the drawing represents 200 mm of the actual object being drawn. This same 1:200 can be thought of as a 1 cm:200 cm or 1 m:200 m, but is expressed simply as a 1:200 ratio.

Table 7–6 shows some typical applications of the metric scale. These applications are roughly analogous to the ones shown in Table 7–5 for the Engineer's scale, and a comparison of these two tables will show similarities and differences between the two scales.

The second example in Table 7–6 points out the necessity of keeping track of decimal places, zeros, and units. Since the metric system is decimal, the size of the numbers can be easily reduced merely by switching units. If the ratio of 1 cm:300 000 000 cm is changed to 1 cm:3 000 000 m, and then to 1 cm:3000 km, the scale is easier to use and less prone to errors.

Table 7–6   Applications of Metric Scale

Scale of Drawing	Scale Selected	Typical Application	Smallest Division on Scale
1:10	1:100	property map	1 cm
1:300 000 000 or (1 cm = 3000 km)	1:300	world map (approx 1″ = 4735 mi)	200 km
1:0.5 (assume units 1 cm = 0.5°C)	1:500	an axis on a graph	0.05°C
1:30 000 or (1 cm = 300 m)	1:300	city map	20 m
1:1	1:100	machine part, full scale	1 mm
1:2	1:200	machine part, half scale	2 mm

EXAMPLE 7–7 Assume a drawing is being made with a scale of 1:25 000, and a distance of 820 m is to be measured. If we temporarily think of the scale as 1 cm = 25 000 cm or 1 cm = 250 m, as shown in Figure 7–61, then the distance of 820 m can be measured without calculation. In order to place the decimal in the correct position, the quantity represented by one cm must first be established.

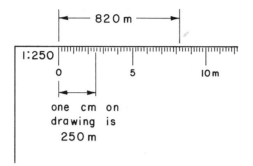

Figure 7–61 Using a metric scale.

A knowledge of the SI (Systéme International d'Unités) or metric system units of length is essential to use of the metric scale. Paper sizes adopted by the International Standards Organization (ISO) are somewhat different from the paper sizes commonly used in the United States. Table 7–7 compares these two systems. Generally the ISO sizes are slightly longer and more narrow than the U.S. sizes.

Table 7–7 Comparison of Paper Sizes

United States Designation	Size of Paper		International Designation	Size of Paper (mm)
	inches	mm		
A	8.5 × 11	216 × 280	A4	210 × 297
B	11 × 17	280 × 432	A3	297 × 420
C	17 × 22	432 × 559	A2	420 × 594
D	22 × 34	559 × 864	A1	594 × 841
E	34 × 44	864 × 1118	A0	841 × 1189

7.9 Dimensions

Although most engineering drawings are made to a specific scale, dimensions also are included. Dimensions provide the reader with the measurements used to produce the part and relieve the reader from the task of measuring the drawing. Moreover, the dimensions control the accuracy to which these lengths are measured. Figure 7–62 shows engineers carefully checking the dimensions of an automobile.

Dimensions are drawn to a set of commonly-agreed-to rules called standards. Many of the standards established by the American National Standards Institute have been reproduced in this book for your convenience. The manner in which dimensions are to be chosen and placed on a drawing have been well established.

Linear dimensions are drawn with a dimension line, broken in the middle for the numerical value, and terminated by an arrowhead on each end. Extension lines show where this dimension appears on the object. Figure 7–63 shows examples of dimension lines, extension lines, a note with leader and shoulder, a general note, and center lines.

Figures 7–64 and 7–65 show the manner in which dimensions are placed on drawings. The larger distances, such as 1.500 and 0.750 in Figure 7–65, are

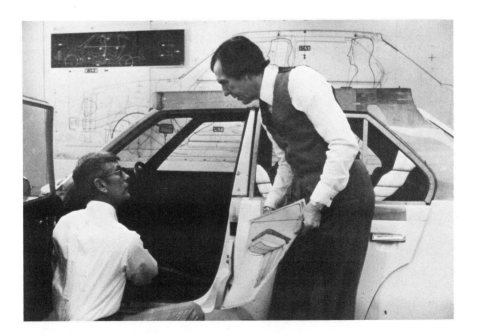

Figure 7–62 Checking dimensions. (*Courtesy* Chrysler.)

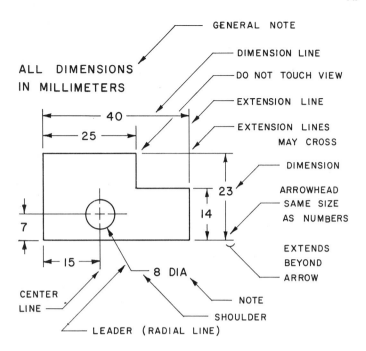

Figure 7-63 Examples of dimensioning terms. (ANSI Y 14.5–1973.)

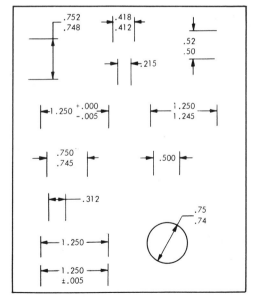

Figure 7-64 Decimal dimensions. (ANSI Y 14.5–1973.)

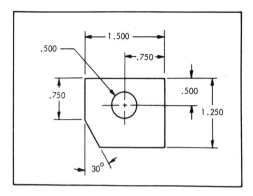

Figure 7–65 Placement of dimensions. (ANSI Y 14.5–1973.)

drawn with the numerical value and the arrowheads inside the extension lines. Smaller dimensions, such as the 0.500 and 30° in Figure 7–64, are placed with the value inside but the arrowheads outside the extension lines to avoid crowding. If the dimension is quite small, both the arrowheads and the value may be moved outside the extension lines, as shown on dimension 0.215 near the top of Figure 7–64. The 0.312 dimension in the same figure shows another method of avoiding crowding. In this case the numerical value is moved outside, but the arrowheads remain inside the extension lines.

Figure 7–66 shows the placement of dimensions for circular shapes. In order to avoid extension lines crossing over dimension lines, dimensions generally are placed with the smaller distances closer to the view and the larger ones on the outside, as shown in Figure 7–67. Figures 7–68 and 7–69 show that arrowheads should terminate at a line rather than at a corner or point. The line terminating the

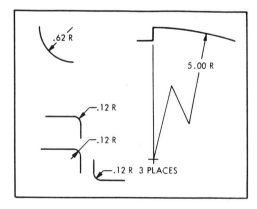

Figure 7–66 Dimensioning Radii. (ANSI Y 14.5–1973.)

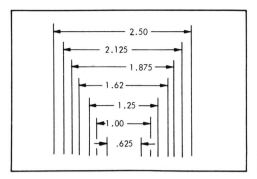

Figure 7–67 Staggered Dimensions. (ANSI Y 14.5–1973.)

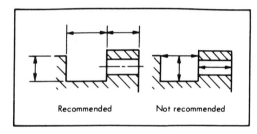

Figure 7–68 Location of dimensions. (ANSI Y 14.5–1973.)

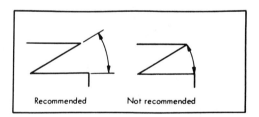

Figure 7–69 Location of dimensions. (ANSI Y 14.5–1973.)

arrowhead may be an extension line or a solid (visible) line. Extension lines may cross each other, as shown in Figure 7–70, but dimension lines should not be crossed either by other dimension lines or extension lines. If extension lines interfere with arrowheads, the extension lines are broken as shown in Figure 7–71. Figure 7–72 shows examples of several varieties of leaders, the long connecting lines between the dimension (or note), and the location of this dimension (or feature) on the drawing. The leaders for notes are drawn at a slant while dimension

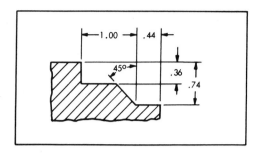

Figure 7–70 Crossing extension lines. (ANSI Y 14.5–1973.)

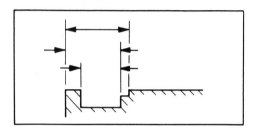

Figure 7–71 Breaks in extension lines. (ANSI Y 14.5–1973.)

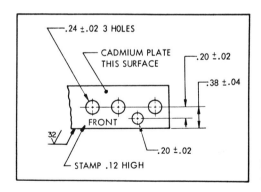

Figure 7–72 Examples of leaders. (ANSI Y 14.5–1973.)

leaders may be shown as vertical or horizontal lines. Leaders pointing to circular shapes always point toward the center, as shown in Figure 7–73.

All dimensions shown in Figures 7–63 to 7–67, 7–70, and 7–72 are unidirectional. This means that all the letters and numbers are placed horizontally and

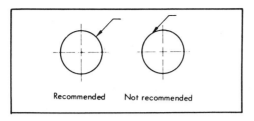

Recommended   Not recommended

Figure 7–73 Leaders. (ANSI Y 14.5–1973.)

may be read from the bottom of the drawing. An aligned system of dimension placement, shown in Figure 7–74, also may be used.

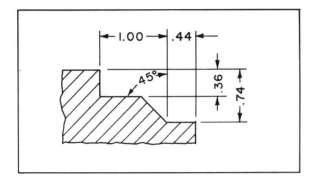

Figure 7–74 Aligned dimensions. (ANSI Y 14.5–1973.)

A brief summary of dimension markings is given in Table 7–8.

Table 7–8 Dimensioning

Name	Description
Dimension line	Thin black line terminated by arrowheads; may not cross
Extension line	Thin black line; does not touch view; extends beyond dimension line; may cross
Arrowheads	Same length as lettering height
Leaders	Used for notes; point toward center of circular shape; have approximately 10 mm shoulders
Center lines	Used to locate center of circular shape; take place of extension line

The selection and placement of dimensions is governed by a set of rules or general guidelines which should be followed whenever possible. These rules may be violated occasionally, but they should not be violated unless there is good reason.

Rule 1. Keep the dimensions off the views. Figure 7–75 shows three possible positions for a dimension. This rule indicates that either A or C is preferable to B.

Rule 2. Avoid putting either too many or too few dimensions on a drawing. Use only those dimensions which are necessary to produce the object being dimensioned. In Figure 7–75 either A *or* B *or* C should be included. In Figure 7–76, one of the four widths can and should be omitted. A, B, and C could be included and D omitted, or A, B, and D included and C omitted. Including only two of these dimensions would give too little information; including four gives more than is needed.

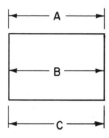

Figure 7–75   Rule 1 for dimensioning.

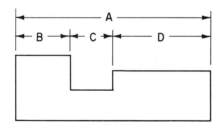

Figure 7–76   Rules 2 and 3 for dimensioning.

Rule 3. Include the overall dimension on the drawing. In Figure 7–76, A should be chosen as one of the dimensions to be included since A indicates the overall dimension.

Rule 4. Avoid dimensioning to hidden lines. In Figure 7–77, dimension U is preferable to R, since U dimensions to a visible rather than a hidden line.

Rule 5. Place the dimension for a feature in the view in which the shape of the feature shows most clearly. In Figure 7–78 dimension S is preferable to U since the feature is more clearly shown in that view.

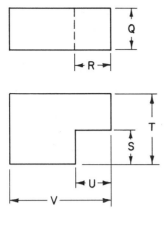

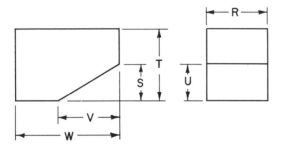

Figure 7–77 Rule 4 for dimensioning.

Figure 7–78 Rule 5 for dimensioning.

7.10 Accuracy of dimensions

No measurement is exact, and consequently all dimensions have limits of accuracy. Limits to the accuracy of a dimension may be viewed from the standpoint of the possible and from the standpoint of the practical. The first consideration, that of how accurately the object *can* be made, is a function of the people and equipment used to make the object. The second consideration, that of how accurately the object *must* be made to function properly, is a function of the design. If the object is made more accurately than is necessary to function correctly, the result is acceptable, and can be used. Since it generally costs more to produce more accurate products, however, it is more profitable to allow larger errors, as long as the errors do not interfere with the quality of the final product.

In Figure 7–79 the hole in part B is dimensioned 11 mm in diameter, and the cylinder on part A is dimensioned 10 mm. These two parts are designed to fit together as shown in Figure 7–80. If we assume that the dimensions in Figure

7–79 are all ± 1 mm, then the possibility arises that the two parts will not fit. This situation could arise if the cylinder on part A had a diameter of (10 + 1 =) 11 mm, and the hole had a diameter of (11 − 1 =) 10 mm. If the dimensions are assumed to be ± 0.1 mm, then the two parts will fit together. This variation in accuracy provides a realistic, workable design that will allow the parts to function.

The variation in accuracy permitted is called the *tolerance* on the dimension. In the preceding case the tolerance is 0.2 mm, or ±0.1 mm. The smaller the tolerance, the greater the cost of manufacturing, so it is undesirable from an

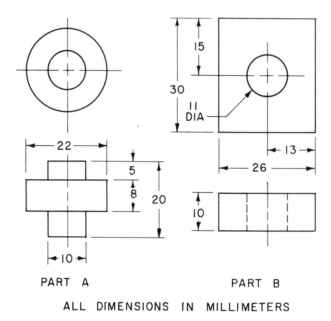

PART A   PART B

ALL DIMENSIONS IN MILLIMETERS

Figure 7–79 Two mating parts, A and B.

economic standpoint to specify an unnecessarily small tolerance. A tolerance of 0.2 mm was found to be satisfactory in the above example, so a tolerance of 0.02 mm would be unnecessary and expensive. In specifying tolerance on a dimension the designer must also consider whether the part can be made as accurately as the drawing requires. If a tolerance of 0.02 mm (11.00 ± .01 mm diameter) were placed on the hole in part B of Figure 7–79, it may happen that the equipment used in creating the hole cannot make it this accurately. If this is the case, then the dimension is realistic from the design standpoint, but not realistic from a production standpoint.

Once the necessary tolerance level has been determined, there are several ways to specify this dimension on the drawing. One common method is to use a

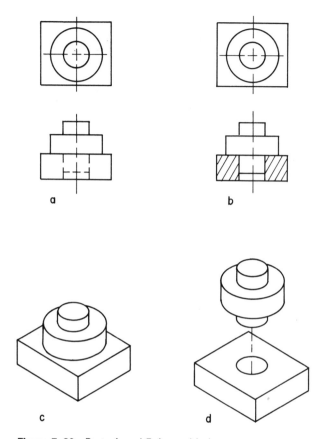

Figure 7–80 Parts A and B Assembled.

general note specifying the tolerance for all dimensions on the drawing, for example, "ALL DIMENSIONS ± 0.5 mm." If some dimensions require more accuracy than others, as is typically the case, then a note may be used which establishes different classes of accuracy:

All dimensions X are ± .5 mm
All dimensions $X.X$ are ± .1 mm
All dimensions $X.XX$ are ± .05 mm

This note tells the reader that a part dimensioned 16.2 mm must be equal to or larger than 16.1 and equal to or smaller than 16.3, or it is not acceptable. These two figures form a *limit dimension* and may be expressed as shown in Figure 7–81a. Alternate means of expressing these same limits are shown in Figure 7–81b and 7–81c. The dimension expressed in view (a) shows no preference for either end of the range, implying any size between 16.1 and 16.3 is equally acceptable. It also requires no calculation by the reader to find the two limits. Figure 6–81b implies that the size of 16.2 is optimum, and an error of 0.1 mm is allowable in either direction. Both views (b) and (c) require a little arithmetic to calculate the two limits, 16.1 and 16.3. The dimension in Figure 6–81c implies that the lower end of the range, 16.1, is optimum, and that the allowable error can only be upward, toward the 16.3 size, for the part to function correctly. For in-

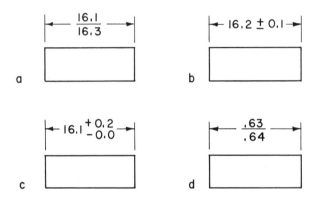

Figure 7–81   Limit dimensions.

spection purposes, all three dimensions provide the same criteria for accepting or rejecting a part.

The accuracy of the limit dimensions required in drawings may vary with the system of units used. As a rule of thumb, for approximately equal accuracy, inch dimensions need one more decimal place than dimensions given in millimeters. For example, the limits of the millimeter dimensions shown in Figure 7–81a convert to 0.63 in. and 0.64 in., as shown in Figure 7–81d. This rule is reversed when converting from inches to millimeters, as shown in Figure 7–82. The 1.730 in. length converts to 43.942 mm which is rounded to 43.94 mm. The tolerance on this dimension $(1.730 - 1.725 = 0.005)$ converts to 0.127 mm which rounds to 0.13 mm. This tolerance $(43.94 - 43.81)$ agrees with view (b).

When two parts fit together, the type of fit which occurs is dependent upon the relationship between the two *mating dimensions*. Figures 7–79 and 7–80 show examples of such mating dimensions. The condition which provides the tightest possible fit of the mating objects is referred to as maximum material condition, and is abbreviated as MMC.

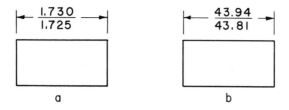

Figure 7–82 Conversion from inches to millimeters.

In Figure 7–83, the tightest fit occurs when the slide is at its maximum and the groove is at its minimum size. At MMC, the *allowance* is +0.4 mm, indicating that air space exists, or that the groove is wider than the slide. The other extreme situation would exist if the largest groove contained the narrowest slide, resulting in an allowance of +0.8 mm.

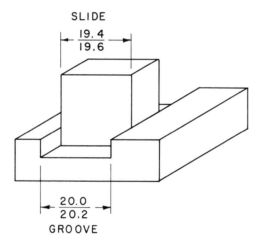

Figure 7–83 Mating part and limit dimensions.

Minimum Allowance (at MMC)	Maximum Allowance
20.0	20.2
− 19.6	− 19.4
0.4	0.8

The loosest fit with +0.8 mm allowance permits the maximum misalignment. If we assume the length of the slide to be 30 mm, the geometry for misalignment will appear as shown in Figure 7–84. Since the angle θ is quite small, we may

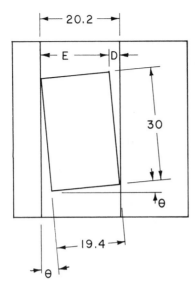

Figure 7–84 Misalignment of mating parts.

assume that E is approximately 19.4, and D is approximately 0.8. The tangent of θ, then, is $D/30$, making θ about 1.53°. To check the error in assuming $E = 19.4$, let $\cos \theta = E/19.4$, and consequently $E = 19.393$. The error is less than 1%. If these two parts can function with a maximum misalignment of $\pm 1.5°$, the dimensions are satisfactory.

7.11 Summary

An engineer must be able to express ideas graphically. This graphical expression may be in the form of a pictorial drawing, a multiview drawing, an assembly drawing, or a detail drawing of a single part such as that shown in Figure 7–85. The drawing may furnish the viewer a quick, easy-to-understand description or may involve great detail. The drawing in Figure 7–85 has enough information to produce this part. To attempt to describe this part without a drawing would be almost impossible. Thus, the ability to express ideas graphically and to understand drawings is vital to a successful designer in the field of engineering.

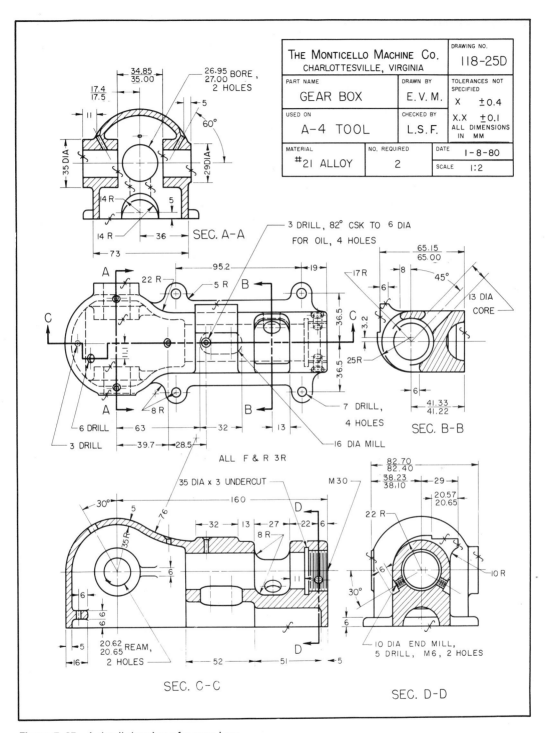

Figure 7–85 A detail drawing of a gear box.

REFERENCES

BEAKLEY, G. C., and E. G. CHILTON, *Introduction to Engineering Design and Graphics*. New York: Macmillan, 1973.

EARLE, J. H., *Engineering Design Graphics,* 3rd ed. Reading, Mass.: Addison-Wesley, 1977.

FRENCH, T. E., and C. J. VIERCK, *Engineering Drawing and Graphic Technology,* 12th ed. New York: McGraw-Hill, 1978.

GIESECKE, F. E., A. MITCHELL, H. C. SPENCER, and I. L. HILL, *Technical Drawing,* 6th ed. New York: Macmillan, 1974.

HAMMOND, R. H., C. P. BUCK, W. B. ROGERS, G. W. WALSH, and H. P. ACKERT, *Engineering Graphics,* 2nd ed. New York: Ronald, 1971.

HOELSCHER, R. P., C. H. SPRINGER, and J. S. DOBROVOLNY, *Graphics for Engineers*. New York: Wiley, 1968.

LEVENS, A. S., and W. CHALK, *Graphics in Engineering Design,* 3rd ed. New York: John Wiley and Sons, 1980.

LEVENS, A. S., *Fundamentals of Engineering Drawing for Design, Communication, and Numerical Control,* 7th ed. Englewood Cliffs, N.J.: Prentice-Hall, 1977.

LUZZADDER, W. J., *Innovative Design with an Introduction to Design Graphics*. Englewood Cliffs, N.J.: Prentice-Hall, 1975.

SCHNEERER, W. F., *Programmed Graphics*. New York: McGraw-Hill, 1967.

Off shore drilling. (*Courtesy* Shell Oil Co.)

Design using standard parts

8

Before 1800 the design of most machines and precision goods assumed that the manufacturing and assembly of these items would be performed by highly skilled craftsmen who would custom-build each unit. This method of design and manufacture produced excellent products, even though each unit was unique and required considerable skill and time to construct. Because of this practice, production rates were low, construction costs were high, and routine repair of equipment was both difficult and expensive. The desire to remove some of the causes of these technological difficulties lead early industrialists to seek new, efficient engineering methods for design and manufacture. Thus in the early 1800s American businessmen, motivated by this desire, began experimenting with new methods of manufacture. One of the first innovations was made by Eli Whitney and a few arms makers from Connecticut. The innovation was an experiment based on the concept of interchangeable parts for guns. This particular effort involved careful control over the design and manufacturing of component parts so that any part would fit any gun. Although the new concept required careful control during the design and manufacturing phases of production, it greatly enhanced the speed and simplicity of assembly since it reduced the need for assembly specialists. In addition, it minimized the previous problems of repair and maintenance. By the time of the War of 1812 these early entrepreneurs had demonstrated that their idea was economically practical. Their efforts were so successful that the process of mass producing standardized articles with interchangeable parts quickly spread to clocks, transportation equipment, farm equipment, and numerous other items. The process of standardization of parts has since developed into our modern mass production technology. As a result, our economy is highly dependent on the availability of standard parts (Figure 8–1).

This chapter will discuss the use of standard parts within the context of modern engineering design. We will first look at the present-day rationale for using standard parts, then at the sources for these standard parts. Finally, we will study the standards that regulate the production of the commonly used standard parts.

As we begin our discussion of standard parts, it is good to keep in mind that the term "standard part" is used in its broadest sense to mean any manufactured commodity that is:

1. carefully designed and manufactured according to universally accepted specifications;
2. produced in large enough quantities to minimize the effect of the cost of design and manufacture on the market price of the commodity; and
3. marketed for use as a component part of a larger engineering system.

Thus, standard parts include not only such obvious things as machine screws,

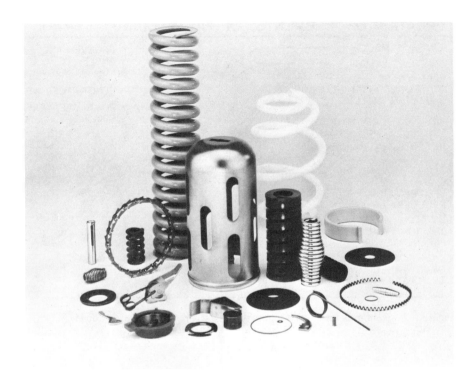

Figure 8–1 An assortment of springs which are available as standard parts. (*Courtesy* Associated Spring, Barnes Group Inc.)

springs, and gears, but also standard materials, such as 1020 steel or neoprene rubber, and standard construction practices, such as welded joints (Figure 8–2).

8.1 The rationale for use of standard parts

There is an old saying that no one likes to re-invent the wheel. This simply means that there is little intellectual reward in trying to re-create something that already exists. Thus, engineers who spend time trying to invent something exactly like a product that is commercially available are wasting their talents and will likely experience the frustration of not having accomplished anything significant. With the present technological needs in the world, there are certainly more exciting areas in which the design engineer could apply his or her energies and talents.

Even if the reasons of interest level and frustration were not present, the economic reasons for using standard parts would still be overwhelming. In mod-

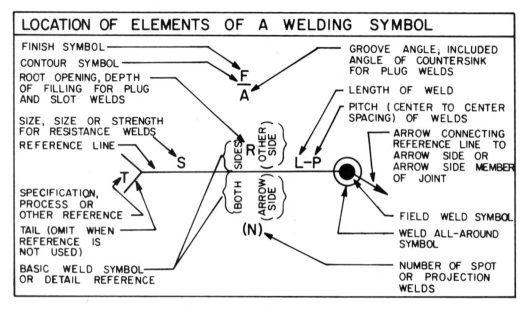

Figure 8–2   Standards for welded joints. (*Courtesy* American Welding Society.)

ern engineering systems the costs of engineering research, engineering design, and manufacturing often are several orders of magnitude greater than the material cost for a single item. Thus, if one can avoid paying these costs or if one can spread these costs over many units, a greater profit can be achieved. As an example, suppose that it is desired to design and manufacture a driven fastener to hold two pieces of lumber together. Even if the engineer spends less than an hour deciding on the shape and size of the body, head, and point of this device, the engineering cost could easily exceed $10. Even if a simple, inexpensive manufacturing method is used, the setup cost associated with the fabrication would likely exceed $20. Thus, one could easily spend in excess of $30 to design and build a single nail which can be purchased at a hardware store for less than one-thousandth of this cost.

Another significant reason for using standard parts is that suitable replacement parts are conveniently available at locations everywhere. Thus, if an engineering student decides to change the spark plugs in a car, the student can be assured that every nearby auto parts store will have the proper type of spark plug in stock. Indeed, it is reasonable to assume that the engineering student will have a variety of brand name spark plugs from which to choose—each of these alternatives meeting the standards of size and performance specified by the manufacturer of the engine.

Another advantage of using standard parts is that the established specifications required for their construction assure the consumer of a product whose performance meets expected standards. In this sense, the consumer is protected from manufacturers of inferior parts.

The widespread acceptance of the metric system in this country now facilitates a greater degree of interchangeability between parts manufactured in this country and parts manufactured elsewhere. Thus, a greater market is now available for the international sale of goods and services.

A final advantage to the engineer who utilizes standard parts in design is that clear description of a part is much easier and faster to achieve. For example the specification "M6 × 1–5e6g" conveys a complete description of the thread type on a particular threaded fastener. This specification together with a simplified symbol can be used on an engineering drawing to convey much information in a very easy way. A standard part represents a complete design module. The drafting symbols for standard parts are easy to draw and are universally recognized.

The widespread use of standard parts is obvious to anyone who studies the construction of any complex engineering system such as an automobile, typewriter, airplane, or building. Over half of the elements in these systems are standard parts which, under other circumstances, could just as easily have found their way into a different system.

The engineer who has a broad understanding of the different available standard parts and their application is a valuable asset to any design team.

Having established the advantages of using standardized parts in engineering design, the next logical topic for discussion is "where does one get information about standard parts?"

8.2 Information on the availability of standard parts

The most obvious place to obtain information about standard parts is from those companies which sell these modular miracles. There are a variety of resources available to the design engineer who wants to identify and locate commercial suppliers of these parts.

The Yellow Pages

One of the best ways to locate suppliers of standard parts is to look in the yellow pages of your telephone directory. Naturally, the more populous your geographic area is, the larger will be your yellow page directory and thus the better will be your chances of finding a local supplier of the particular item you need. Even if the manufacturer is located far from where you work, you may still find help through the yellow pages of your telephone book since suppliers of standard parts frequently have nationwide toll-free telephone numbers for quick answers to specific questions about standard parts. In addition, companies that market on a nationwide basis often have local field representatives who live and work in your geographic area. These specialists are knowledgeable about the advantages and limitations of their products and are often willing to accept your invitation for ''on-site'' advice concerning the application of their standard parts (Figure 8–3). Because of his or her technical training and professional approach, the sales engineer/field representative is usually regarded as a valuable consultant to the design process and is seldom thought of as a high-pressure door-to-door salesperson. In addition to providing technical information, field representatives can also give information on price and delivery time on orders for standard parts. Such information is a necessity to the engineer who has both time and economic constraints.

Design libraries

Most companies involved in engineering design have their own special-purpose libraries containing catalogs and other descriptive literature about the avail-

Figure 8–3 A field service representative of a company that manufactures springs discusses their use with a client. (*Courtesy* Associated Spring, Barnes Group Inc.)

ability and application of standard parts. Suppliers gladly provide free literature about their products, and some design engineers find it helpful to maintain their own personal libraries of frequently used information on standard parts. Since part prices can change more rapidly than part specifications, most standard part suppliers publish their price lists separately from their descriptive catalogs. Thus, when requesting catalog information from suppliers, it is wise to request a list of current prices. When telephoning or writing to a supplier located some distance from your location it is also wise to inquire whether the company has a local representative in your area. Because they are arranged to be of the maximum use to design engineers, manufacturers' catalogs are clearly illustrated, are easily understood, and often contain extra technical information of value to the design engineer (Figure 8–4). The volume of catalog material concerning standard parts is vast. A few companies have exploited this fact and are in the business of providing complete libraries of information that include regular updating service to their customers. Some of these special libraries use sophisticated information retrieval equipment such as microfilm files and computer search terminals to enhance their efficiency.

Trade publications

One of the important characteristics of a profession is that its members engage in self-study to keep abreast of new developments. So it is with the engi-

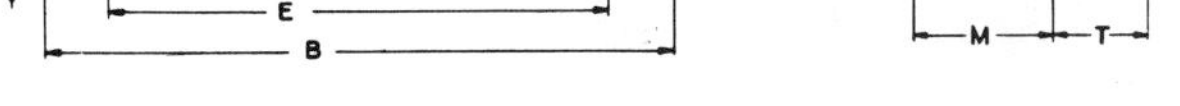

Series SRP

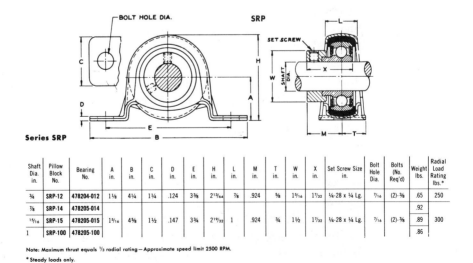

Shaft Dia. in.	Pillow Block No.	Bearing No.	A in.	B in.	C in.	D in.	E in.	H in.	L in.	M in.	T in.	W in.	X in.	Set Screw Size in.	Bolt Hole Dia.	Bolts (No. Req'd)	Weight lbs.	Radial Load Rating lbs.*
¾	**SRP-12**	**478204-012**	1⅛	4¼	1¼	.124	3⅜	2¹³⁄₆₄	⅞	.924	⅝	1⁵⁄₁₆	1⁷⁄₃₂	¼-28 x ¼ Lg.	⁷⁄₁₆	(2)-⅜	.65	250
⅞	**SRP-14**	**478205-014**															.92	
¹⁵⁄₁₆	**SRP-15**	**478205-015**	1⁵⁄₁₆	4⅝	1½	.147	3¾	2¹⁹⁄₃₂	1	.924	¾	1½	1⁷⁄₃₂	¼-28 x ¼ Lg.	⁷⁄₁₆	(2)-⅜	.89	300
1	**SRP-100**	**478205-100**															.86	

Note: Maximum thrust equals ⅓ radial rating—Approximate speed limit 2500 RPM.

*** Steady loads only.**

Figure 8–4 Manufacturers' catalogs are clearly illustrated, are easily understood, and contain useful technical information. (*Courtesy* SKF Industries, Inc.)

neering designer. Many trade publications are available to help the design engineer with this task. The trade publications of interest to engineering designers fall into three categories.

The first category of publication contains feature articles and advertisements that cater to disciplinary groups. This means that they are written for chemical engineers, mechanical engineers, electrical engineers, etc. These publications are often published by professional societies and frequently are provided along with membership. Good examples of these are *Chemical & Engineering News* published by the American Chemical Society and *Mechanical Engineering* published by the American Society of Mechanical Engineers.

The second category of trade publication cuts across disciplinary boundaries and is oriented toward job function. This means that they are written for use by design engineers, manufacturing engineers, sales engineers, etc. Examples of this type of publication are *Product Engineering* (for design engineers) and *Metalworking Today* (for manufacturing engineers).

The third type of trade publication is oriented toward a specific product or industry. For example, *Aviation Today* is a trade magazine that serves the aircraft industry, *Automotive Engineering* is a magazine that serves the automotive industry, and *Power* is a magazine that serves the needs of engineers in the power industries.

An excellent categorical listing of trade journals is presented in the *Annual Listing of Corporations and Industries* by Funk & Scott. This listing is a helpful resource for design engineers wishing to find trade publications in their areas of specialization. Because they are easy to read, include feature articles, and contain generous amounts of advertising, these publications are extremely useful to the design engineer. Since a large portion of the income of these publications comes through advertising fees, subscriptions to many of these publications are free to qualified engineers. Almost without exception, these publications contain reader service cards that can be used to obtain further information about the products and standard parts advertised in their issues (Figure 8–5).

Special directories

To do an optimum job of shopping for standard parts, the engineer needs to know the names of different companies that competitively market the same or similar products. To assist the engineer with this process a number of special directories are available. These fall into three useful categories.

The first category contains an alphabetical listing of products and services in a given area of engineering. The design engineer might use this directory to locate all companies that market a particular product or service. This type of directory lists the name, address, and telephone number of each company, and the

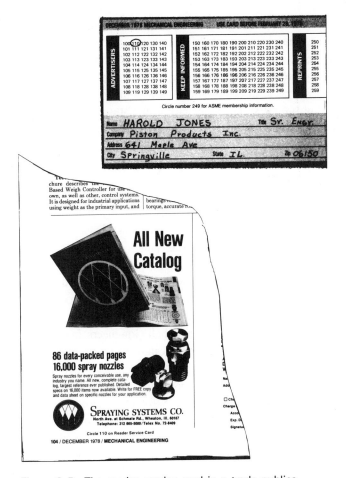

Figure 8–5 The reader service card in a trade publication allows the user to obtain further information about the products or services advertised. (*Courtesy Mechanical Engineering* and Spraying Systems Co.)

engineering designer will often write to each of the companies to request their latest catalog and price list. Sometimes this type of directory is published by a trade magazine in the form of an annual buyers' guide.

The second type of special directory lists names, addresses, and telephone numbers of companies that conduct business in the United States. These directories are frequently alphabetical by company name and are thus useful if one wishes to locate a company from the nameplate on a particular piece of equipment. A good example of this type of design directory is the *Thomas Register of American Manufacturers*. This special-purpose register also gives information on the size and capital rating of the companies it lists. This information is of interest because it gives an indication of the production capacity and financial stability of each listed company.

The last type of special-purpose directory contains abbreviated catalogs of various companies arranged so that similar catalogs of similar products are side-by-side. This type of directory allows easy comparison of the relative advantages of competing products and is actually a miniature design library contained in a table-top form. A good example is the multivolume directory *Sweet's Catalog File—Products for Engineering* published by McGraw-Hill.

Trade shows

The dream of every shopper is that all stores that sell similar items would be located side-by-side within a single city block. This, of course, is a dream that seldom comes true for the average consumer. It does happen, however, for the engineering designer in the form of trade shows. During these annual events, suppliers of standard parts set up displays in rented booths and staff their booths with their best engineering sales personnel. The booths are stocked with free catalogs and also display the products of the company in a tasteful, attractive fashion. The booths are contained in an exposition hall. Engineers frequently travel long distances to attend these events and obtain information to assist them with their design tasks. Since the trade shows last for several days, the engineers who visit these events usually have ample opportunity to visit with the many suppliers who have their products displayed (Figure 8–6). On entering the exposition hall, the visiting engineer is often provided with a special embossed card that lists the name and address of his or her company. The card resembles a credit card, and each exhibitor is provided with a machine that will make impressions of the information it contains onto the forms of the exhibitor. This is an easy way for the visiting design engineer to request information and catalogs by mail.

Various national expositions on special areas of interest to engineering design are held annually. Most of these are well attended. One of the largest of these is the Design Engineering Show sponsored each spring by the American Society of Mechanical Engineers. This show draws exhibitors and visitors from all over the world.

8.3 Recognized standards

With so many standard parts being used in design practice, two questions come to mind.

1. Who sets the standards?

2. Where can information be obtained about these standards?

Figure 8–6 Trade shows provide an opportunity for design engineers to talk with representatives of manufacturers and suppliers of standard parts. (*Courtesy* Clapp and Poliak, Inc.)

The most logical groups to set standards are the two closest to the products—the manufacturers and the users. Engineers from the user community have a strong interest in seeing that the standard specifications for parts are written in a way that will provide maximum utility. In like manner, engineers from the manufacturing companies have a strong economic interest in seeing that their products have the widest possible use in the engineering market place. For this reason, national committees of engineers are often brought together under the auspices of engineering professional societies or trade organizations to establish and to document national standards that are acceptable to all sectors. The national standards system is presently administered by ANSI (The American National Standards Institute) in New York. Companies may subscribe to membership in ANSI and may purchase copies of standards from this organization. Some of the sponsoring organizations of ANSI are listed in Table 8–1. ANSI provides a valuable service

Table 8–1 Sponsor organizations that participate in the preparation of standards*

Architectural Aluminum Manufacturers Association
Association of American Railroads
American Association of State Highway Officials
American Association of Textile Chemists and Colorists
American Concrete Institute
American Dental Association
Anti-Friction Bearing Manufacturers Association, Inc.
American Gear Manufacturers Association
Association of Home Appliance Manufacturers
American Nuclear Society
American Petroleum Institute
Air-Conditioning and Refrigeration Institute
Acoustical Society of America
American Society of Agricultural Engineers
American Society of Heating, Refrigerating and Air-Conditioning Engineers
American Society of Lubricating Engineers
American Society of Mechanical Engineers
American Society for Quality Control
American Society of Sanitary Engineers
American Society for Testing and Materials
American Welding Society
American Water Works Association
Builders Hardware Manufacturers Association
Conveyor Equipment Manufacturers Association
Compressed Gas Association
Diamond Core Drill Manufacturers Association
Electronic Industries Association (formerly Radio-Electronics-Television Manufacturers Association)
Fluid Controls Institute
Federation of Societies for Paint Technology
Institute of Electrical and Electronics Engineers (formerly Institute of Radio Engineers)
Illuminating Engineering Society
Institute of Makers of Explosives
Institute of Petroleum (London)
Institute of Printed Circuits
Insulated Power Cable Engineers Association
Instrument Society of America
International Organization for Standardization
Institute of Traffic Engineers
Mechanical Power Transmission Association
National Association of Architectural Metal Manufacturers
National Bureau of Standards
National Committee for Clinical Laboratory Standards
National Electrical Manufacturers Association
National Fire Protection Association
National Fluid Power Association

National Lubricating Grease Institute
National Microfilm Association
National Woodwork Manufacturers Association
Recreational Industry Vehicle Association
Resistance Welder Manufacturers Association
Society of Automotive Engineers
Scientific Apparatus Makers Association
Swedish Standards Association
Screen Manufacturers Association
Simplified Practice Recommendation
Technical Association of the Pulp and Paper Industry
The Fertilizer Institute
Underwriters Laboratories

*Table courtesy ANSI.

both for the engineering community and for the consuming public. ANSI publishes a *Catalog of American National Standards* which alphabetically lists the titles of available standards.

8.4 Threaded fasteners

Because of its extreme versatility and utility, the threaded fastener is perhaps the most frequently used standard part in engineering practice. The basic terminology for screw threads is shown in Figure 8–7. Prior to 1841 each manufacturer of threaded fasteners designed threads according to individual preference. Some of the designs were quite good and others performed poorly. For example, fasteners with threads in the form of a sharp ''V'' were easily broken since the sharp crests were weak and the sharp valleys had high stress concentrations. In order to overcome these disadvantages, in 1841 Sir Joseph Whitworth proposed a thread shape with round crests and roots having an angle of 55°. The Whitworth thread became the British standard. In 1865 in the U.S., William Sellers proposed a thread shape with flat roots and crests and a 60° thread angle. This thread design lead the way to the national and international standards in common practice today. We will now look at the two thread standards that are in present use.

Unified American Screw Threads

The Unified American Screw Thread series contains six different series standards.

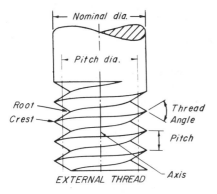

EXTERNAL THREAD

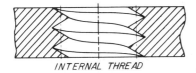

INTERNAL THREAD

Figure 8–7 Basic terminology of screw threads.

The *coarse-thread series* is designated "UNC." This class is recommended for general use and has the fewest number of threads per inch of any of the series.

The *fine-thread series* is designated "UNF." This class is recommended where extra strength and vibration resistance are important. This class is frequently used in automotive and aircraft applications.

The *extra–fine-thread series* is designated "UNEF" and is used for applications requiring a maximum number of threads per inch for a given diameter.

The *8-thread series* is designated "8N" and is a series having the same pitch for all diameters. This series is often used to join the flanges of high pressure pipe.

The *12-thread series* is designated "12UN" and often is used to provide extra fine threads for diameters of 1.5 in. or larger requiring fine threads.

The *16-thread series* is designated "16UN" and often is used to provide extra fine threads for diameters of 2 in. or greater.

In addition to categorizing threads according to pitch series, threads are also classed according to the amounts of tolerance they exhibit. Class designations "A" are for external threads and designations "B" are for internal threads. As a general rule, the higher the class number, the lower will be the allowance and tolerance and thus the higher will be the quality of the fastener.

Classes 1A and 1B are intended for general use where free assembly and easy production are desired. Classes 2A and 2B are the recognized standard for most screws, bolts, and nuts. Classes 3A and 3B represent exceptional quality, commercially threaded fasteners.

Although screw threads are helical as indicated in Figure 8–7, it is extremely cumbersome to draw them as they actually appear. There is, of course, little need to draw an exact picture of screw threads on an engineering drawing since the standard thread specification contains a complete description of the threads. For this reason simplified symbols for threads are used. Accepted standard symbols for external threads of diameter less than 1 in. are shown in Figure 8–8. Accepted standard symbols for internal threads of diameter less than 1 in. are shown in Figure 8–9. For thread sizes over 1 in. in nominal diameter it is customary to draw a semiconventional thread representation as shown in Figures 8–10 and 8–11. The specification information required to completely define the thread on a fastener includes the nominal diameter, the pitch (threads per inch), and thread class. If the thread is left hand, the letters LH are included in the specification.

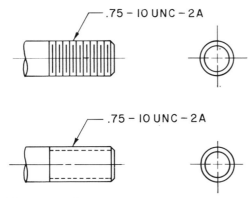

Figure 8–8 Accepted standard symbols for external threads of diameter less than 1 in.; Unified American System specifications.

ISO metric threads

The International Organization for Standardization (ISO) was organized in 1946 to establish a single, international standard for screw threads. The result of their work is a metric thread system that is rapidly becoming the universal thread system of the world. The diameter–pitch combinations in this system are presented in table form in the appendix of this book. A complete description of the threads in this system involves specifying the nominal size in millimeters, the pitch, and the tolerance grades for both pitch crest and pitch diameters. The specification would take the form:

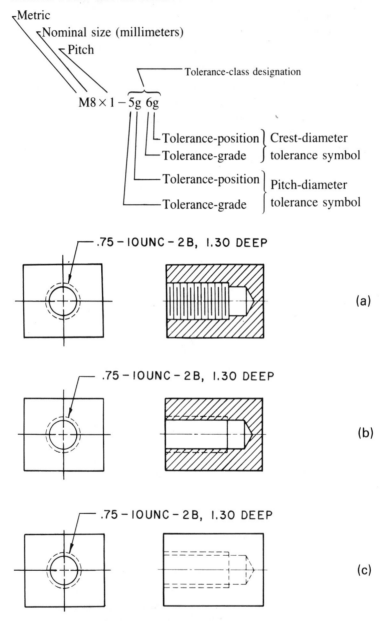

Figure 8-9 Accepted standard symbols for internal threads of diameter less than 1 in.; (a) schematic symbol for section view, (b) simplified symbol for section view, and (c) exterior view; Unified American System specifications.

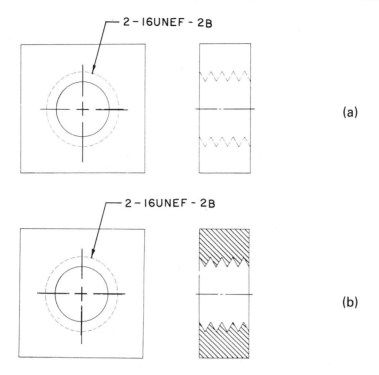

2 — 16UNEF - 2B

(a)

2 — 16UNEF - 2B

(b)

Figure 8–10 Accepted standard symbols for internal threads of diameter 1 in. or larger: (a) external view, (b) section view; Unified American System specifications.

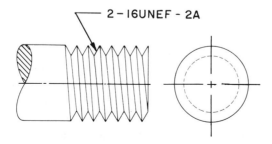

2 — 16UNEF - 2A

Figure 8–11 Accepted standard symbols for external threads of diameter 1 in. or larger; Unified American System specifications.

Examples of the use of this specification are presented in Figures 8–12 and 8–13. If the pitch diameter and crest diameter tolerance symbols are the same, the symbol is given only once in the specification statement. The tolerance grades for this system are as follows:

For external threads:
Lowercase e—large allowance
Lowercase g—small allowance
Lowercase h—no allowance

For internal threads:
Uppercase G—small allowance
Uppercase H—no allowance.

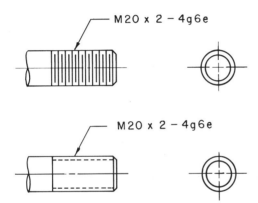

Figure 8–12 Accepted standard symbols for external threads of diameter less than 25 mm; ISO metric system specifications.

8.5 Other standard parts

Although threaded fasteners are the most widely used standard part, the array of standard parts for other engineering applications is nearly endless. The types of parts that an engineer will most often specify will depend on the area of disciplinary training and on the particular industry in which he or she works.

For example, to an electrical engineer the use of standard parts would mean the application of electronic components such as resistors, capacitors, semiconductors, chips, switches, connectors, and transducers. A visit to an electronic parts store would give us a first-hand exposure to the many variations of standard electrical parts.

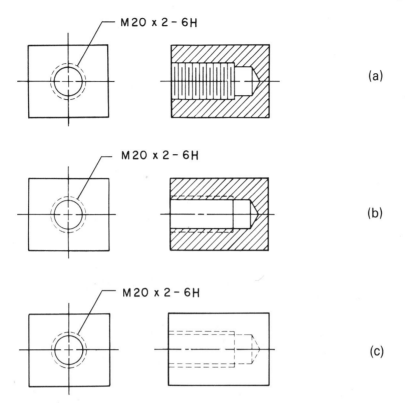

Figure 8–13 Accepted standard symbols for internal threads of diameter less than 25 mm; (a) schematic symbol for section view, (b) simplified symbol for section view, and (c) exterior view; ISO metric system specifications.

To the chemical engineer the use of standard parts would mean the application of commercially available hardware for processing and transporting chemical materials and products. It might also include commercially available chemical materials in liquid, solid, powder, or gaseous form. A visit to a chemical supply house would provide opportunity to observe the many different types of standard parts available for chemical engineering applications.

To the civil engineer the use of standard parts would mean the application of commercially available construction materials including different types of concrete, and different types of structural elements such as I-beams, girders, and reinforcing rods. A careful inspection of the construction elements of a large building or bridge will reveal the many standard parts that are used in civil engineering to provide low cost and high quality.

To the mechanical engineer the use of standard parts would include fasteners, gears, chains, bearings, cables, and springs, as well as standard equipment and materials from the energy-related industries. A visit to the local hardware store would give us a partial glimpse of the many useful standard parts that are available as mechanical fasteners alone.

Just as each engineering discipline has its own unique array of standard parts, so also do particular industries. For example, in the petrochemical industry there are numerous standard parts such as valves, gauges, pipes and fittings, drill bits, and well casing materials (see Figure 8–14). In the automotive industry there are countless standard parts such as tires, batteries, lights, spark plugs, etc. In the aircraft industry there are numerous standard fasteners, component parts, and electronic subassemblies that make design and manufacturing easier.

Regardless of the discipline or industry, the use of standard parts nearly always means better products and lower costs—two characteristics that benefit both the producer and the consumer.

Figure 8–14 Pumping and metering systems at a petro-chemical facility. (*Courtesy* Shell Oil Co.)

8.6 Summary

The ability to make effective use of standard parts in design is a talent that every engineer should cultivate. Knowing when it is appropriate to use a standard part, knowing what type of standard part is best to use, and knowing where to obtain the part offer an exciting challenge to every designer. The increased use of standardized parts and the development of better, international part specifications has brought our world much closer together. In the spirit of international cooperation and in the economic interest of all concerned, this trend will likely continue in the future.

8.1 Use the yellow pages of your telephone directory to locate local suppliers of the following items:

(a) hydraulic cylinders   (f) speed reducers

(b) compression springs   (g) titanium

(c) integrated circuit components   (h) welding rods

(d) ball bearings   (i) methanol

(e) ball bushings   (j) finned tubing

8.2 What is the difference between a technical journal and a trade magazine? Can you find an example of each of these that a civil engineer might use?

8.3 Name three ways that an engineer can become better acquainted with new types of standard parts.

8.4 Locate a copy of Thomas' Register in your library and look up the names and addresses of at least one company that markets each of the following items:

(a) solar collectors   (c) artificial human hip joints

(b) zirconium   (d) sound absorbing foam

8.5 State three advantages and three disadvantages to the use of standard parts.

8.6 Under what conditions would it *not* be advisable to use a standard thread on a threaded fastener?

8.7 Go to your library and see if you can find a trade publication that would be of interest to noise control engineers.

8.8 Suppose that you work for a company that manufactures electric typewriters. What reasons would you give to your supervisor for wanting to attend the next annual ASME Design Show?

8.9 How would you go about finding the date and location of the next annual ASME Design Engineering Show.

8.10 Why do you think that standard parts were not used prior to the 1800's?

8.11 What are the advantages to the use of a reader service card in the back of a trade publication?

8.12 What type of college degree would be best for a local representative of a part manufacturer to have?

8.13 If you were going to start a trade publication of interest to engineers in the gas pipeline industry, how would you go about identifying potential advertisers for your product?

8.14 A certain electronic circuit board has a special connector attached to it. Since the connector has been damaged through improper use, it is your task to find an exact replacement as soon as possible. The connector has the word "AMP" embossed on its side. Can you use this information to find the name, address, and telephone number of the manufacturer?

8.15 What organizations are involved in setting standards for the design and use of the following items:

(a) pressure vessels

(b) elevators for public use

(c) small electric appliances

(d) automobile tires

8.16 What ISO Metric threads are closest in diameter and pitch to the following Unified threads:

(a) ¾-10UNC-2A

(b) ¼-20UNC-3B

(c) 2-16UNF-2B

(d) 1-8N-2A

(e) 1-16UN-3A

REFERENCES

American National Standards Institute (ANSI), "Graphic Symbols for Electronics Diagrams," Y32.2-1967. New York, 1968.

American National Standards Institute (ANSI), "Graphic Symbols for Fluid Power Diagrams," Y32.10-1967. New York, 1967.

American National Standards Institute (ANSI), "Graphic Symbols for Process Flow Diagrams in the Petroleum and Chemical Industries," Y32.11-1961. New York, 1961.

ARNELL, A., *Standard Graphical Symbols: A Comprehensive Guide for Use in Industry, Engineering, and Science*. New York: McGraw-Hill Book Company, 1963.

DREYFUSS, H., *Symbol Sourcebook—An Authoritative Guide to International Graphic Symbols*. New York: McGraw-Hill Book Company, 1972.

Engineers Joint Council, *Thesaurus of Engineering and Scientific Terms*, 1st ed. New York, 1967.

F&S Index of Corporations and Industries, Predicasts, Inc., Cleveland, Ohio.

Institute of Electrical and Electronics Engineers (IEEE) and American National Standards Institute (ANSI) "Graphic Symbols for Electrical and Electronics Diagrams" (IEEE No. 315, March 1971, ANSI Y32.2-1970). New York, 1970.

POLON, D. D., *Encyclopedia of Engineering Signs and Symbols (EESS)*, The Odyssey Scientific Library. New York: The Odyssey Press, 1965.

SIMONTON, D. P. and J. T. MILEK, "Condensed Engineering Language: A Bibliography," *Standards Engineering*, April/May, 1969, pp. 4–9.

1979 Sweet's Catalog File—Products for Engineering, Mechanical and Related Products, Electrical and Related Products, Civil and Related Products. McGraw-Hill Information Systems, Inc., Sweet's Division, New York: 1979.

Thomas Register (Volumes 1–7 Products and Services Listing; Volume 8 Company Listings; Volumes 9–14 Manufacturers' Catalogs), published yearly, New York.

Calvert cliffs nuclear plant. (*Courtesy* Bechtel Power Corporation and Baltimore Gas and Electric Company.)

Systems design and project planning

9

The systems approach to problem solving is a design method well suited for solving large-scale problems involving complex interactions among the various elements in the system. The systems design methodology had its beginnings during World War II when there was a strong need for better managerial and operations techniques. The methods and ideas initially developed over 30 years ago have since been improved and refined into concepts and techniques that are extremely powerful and capable of being applied to a wide variety of problems in science, technology, business, and the social sciences.

Good planning and careful management of an engineering project can often mean the difference between economic success and failure. For this reason every engineer should have a basic understanding of the techniques available for the efficient management of engineering projects. This becomes especially important for projects involving many subsystems and many persons (Figure 9–1).

A thorough description of the methods for project planning or the theory of systems analysis and design would require a deeper treatment than is possible here. For this reason, it is intended to use this chapter to provide an overview of some of the key features of these powerful engineering tools and to illustrate how they can be used to good advantage in engineering problem solving. The reader interested in an in-depth study of these topics should consult the references at the end of this chapter.

9.1   When should the systems design approach be used?

The traditional design method outlined in Chapter 1 is well suited to the solution of succinct problems in which the desired goal is to perform an optimum design of a single product or simple process. As the design methodology is applied to more complex devices and processes it becomes necessary to break the problem down into subsystems. For example, in the design of a car one might consider the design of frame, body, and power plant as three separate sub–tasks in the design process. The traditional design method is adequate for this design situation so long as there are minimal interactions between the subsystems. When this condition is not met, the traditional design approach fails to perform well. The larger a system becomes in complexity the more likely it will be that an optimization of all subsystems will not give an overall optimum for the total system. In fact, it can be shown that complex systems exhibit characteristics in which optimization of a subsystem tends to produce undesirable effects on the performance of the total system. As an example one can consider the design of the automobile as a component in an urban transportation system. The optimum design of the automobile as a subsystem has produced a number of undesirable effects on the overall system such as increased traffic congestion, increased air pollution, higher accident rate, higher energy consumption, etc. Thus it is that when systems become complex to the point that subsystems have strong interactions, a new design approach called the "systems design method" should be used. This method places emphasis on a consideration of optimum design of the system as a whole. Thus the size and complexity of a design problem are important to consider in choosing between the systems design method and the traditional design approach.

Another important factor to consider in choosing the approach to use in design problems is the predictability of the variables involved. The traditional design method is best suited for problems that are "deterministic." This means that every change in the input of a system produces a unique, predictable response of the output. Deterministic systems produce the same output every time the same input is applied. For example the gears in the transmission of an automobile move

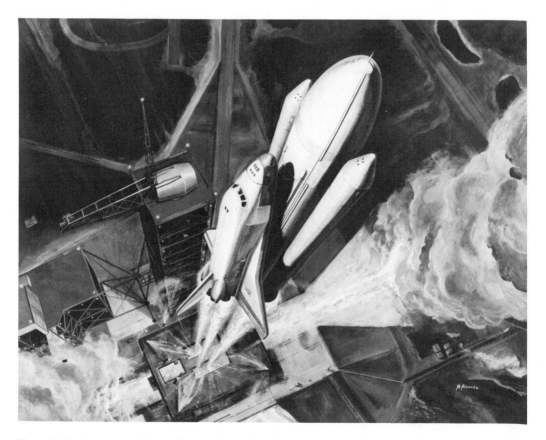

Figure 9–1   Large, complex projects such as the space shuttle would not be possible without the use of systems design and efficient project management. (*Courtesy* NASA.)

in a complex but predictable way so that if a change is made in the number of teeth on one of the gears, the effect can be determined with complete certainty. The systems design method, on the other hand, is often better suited for problems involving "stochastic" systems. Stochastic systems may involve random behavior. In such systems the output response to a given set of inputs may be different each time the same inputs are applied. An example of a stochastic system is the flow of traffic at a busy freeway intersection. Inputs to this system include the weather conditions, the skill of the drivers involved, the type of vehicles in the traffic, the time of day, etc. In this system the traffic flow will be greatly influenced by the incidence of an accident; yet the occurrence of this event cannot be predicted with certainty based on the input conditions.

A third factor to consider in choosing between the traditional design method and the systems approach is whether the solution process involves a large number of people interacting as a group to produce the output. When this situation occurs, the design is best accomplished by the systems approach since it gives a common framework for the thought process that must take place. The need for such a thought framework is especially important when the project involves people from diverse disciplinary areas. Thus, the systems approach provides a common language for communication among such groups as engineers, physical scientists, social scientists, behavioral scientists, lawyers, and medical personnel.

Armed with a knowledge of when to use the systems approach, we are now in a position to see how it works.

9.2   What is the systems approach and how does it work?

A number of ways of describing the systems approach are in common use. Although each of these uses a different visual model, each describes the same method. The formulation shown in Figure 9–2 is one of the better known visual descriptions of the systems design approach. This method has four steps.

Step 1: Defining the objective: In this step the overall objective of the design process is stated in a succinct manner. Sometimes a designer will be given this statement, and at other times he or she will need to provide this definition. Care used in preparing this definition can often save considerable time later in the design process.

Step 2: Requirements: In this step the requirements needed to meet the objective are established and are stated in the most general form possible.

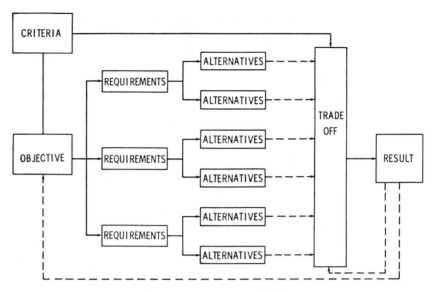

Figure 9–2   The systems design methodology. (NASA
CR 120338.)

Step 3: Alternatives: In this step a list is made of all possible alternatives that
will satisfy each of the requirements.

Step 4: Trade-off: In this step a set of criteria and constraints are used to choose
the combination of alternatives that best meets the objectives.

One of the interesting features of this description of the systems design
method is that it allows each of the steps described above to be considered as a
separate subsystem to which the systems approach can also be applied. Thus, to
obtain a statement of the objective for example, the systems approach diagram of
Figure 9–2 can be applied. In this situation the objective would be to define the
objective for the primary system study. The designer would need to establish the
requirements necessary to define this objective. The output of this subsystem design cycle would be the objective statement for the primary study. If it is necessary, the steps in the subsystem could be further broken down into sub-subsys-
tems. This nested approach to increasing the resolution of the design process can
be carried to as many nesting levels as are necessary to make each step tractable.
After the trade-off for the primary system has been completed and the final
result emerges, it may become obvious that the result is not what is really desired.
In this case it may be necessary to return to the starting step of the design cycle,

modify the objective statement, and repeat the whole process. In this sense the systems approach is often iterative.

Perhaps the best way to illustrate the systems approach is by means of an example.

9.3 An example of the system design method

During the summer of 1974 a group of scientists, engineers, and social scientists came together at the NASA George C. Marshall Space Flight Center in Huntsville, Alabama, for the purpose of studying the problems and challenges of future energy consumption in the United States. The group was known as the MEGA-STAR team—an acronym for "The *M*eaning of *E*nergy *G*rowth: an *A*ssessment of *S*ystems, *T*echnologies, *a*nd *R*equirements." Project MEGASTAR was jointly sponsored by NASA and the American Society for Engineering Education and was managed by Auburn University. The MEGASTAR team recognized that energy policy decisions made in the late 1970's and early 1980's could have significant impacts on our energy consumption patterns and on our way of life in the future (see Figure 9–3). The team also recognized a need to know what the influ-

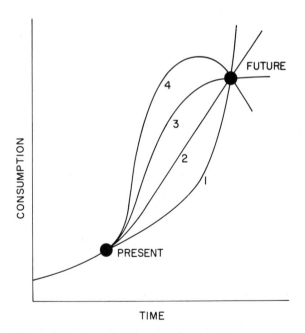

Figure 9–3 Several different estimates of energy consumption for the future are proposed as a basis for formulating energy policy decisions. (NASA CR 120338.)

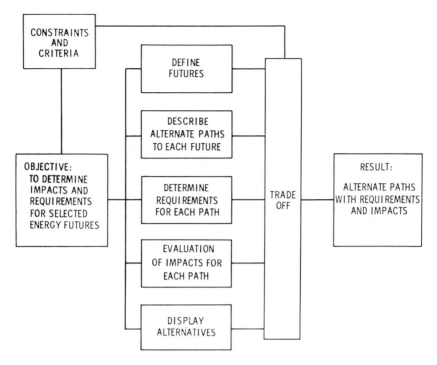

Figure 9–4 The MEGASTAR systems design approach.
(NASA CR 120338.)

ence of any given energy path would be on our way of life for the future. Since the time for the study was limited to eleven weeks, since the system involved was extremely large with complex interactions, since the variables involved were stochastic, and since the men and women in the team were from a diversity of disciplinary backgrounds, the systems approach was clearly the most efficient mode of operation. The MEGASTAR project provides an excellent example of how the systems design approach is applied.

The initial procedure for implementing the systems design approach in the MEGASTAR study is shown in Figure 9–4. In this system diagram the requirements and alternative steps have been combined into the five elemental blocks shown. These blocks were in turn broken into subsystems in order to allow more detailed treatment. For example, in order to determine the requirements for each

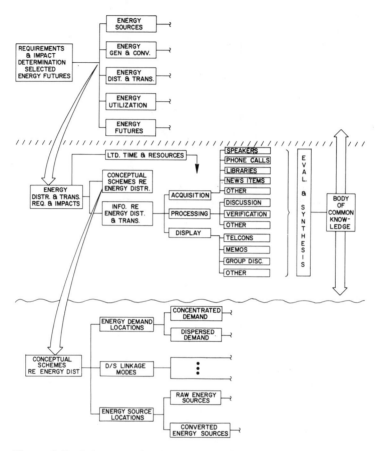

Figure 9–5 Subsystem breakdown used for project MEGASTAR. (NASA CR 120338.)

path, the subsystem breakdown shown in Figure 9–5 was used. As is obvious from this figure, further levels of subsystem breakdown were used.

The conclusions and impacts of the MEGASTAR project were many. Among them were predictions of energy consumption for several paths to the future (see Figure 9–6). In this figure path 1 indicates our energy future if our present rate of energy consumption is allowed to grow unchecked. Path 3 represents what might be possible if we use both conservation and technological improvements in our generation and use of energy. Path 2 represents what the future might look like if we take technological steps to reach zero energy growth by the year 2000. The resource requirements for each of these energy paths to the future were investigated, and the results are quite interesting. For example the MEGA-STAR final report looks carefully at the engineering manpower requirements (see Figure 9–7) and capital expenditure requirements (see Figure 9–8) associated with different energy paths. The detailed adding together of dollars and engineers required to implement each energy path would not have been possible had it not been for the detailed breakdown of the system being considered. This breakdown is one of the strongest advantages to using the systems design approach.

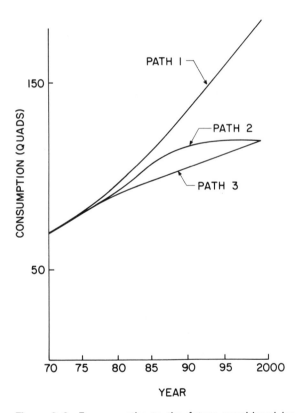

Figure 9–6 Energy paths to the future considered by project MEGASTAR. Path 1 represents continued, exponential energy growth; Path 2 represents conservation sufficient to arrive at the year 2000 with zero energy growth rate; and Path 3 represents severe energy conservation. (NASA CR 120338.)

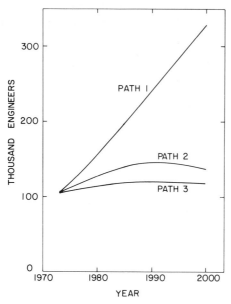

Figure 9–7 Engineering manpower requirements for various energy futures. (NASA CR 120338.)

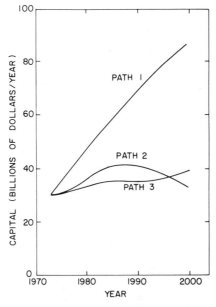

Figure 9–8 Capital requirements to implement various energy futures. (NASA CR 120338.)

9.4   Project planning and project management

Once a design concept has been established for the solution of a given project objective, the next logical step is to see that the project is completed through an implementation of the design in an efficient manner. The engineer frequently serves as manager of a project. It is the purpose of this section to discuss methods that are available to help the project manager to perform the intended task in an efficient manner.

The Gantt chart

Before 1900, project management was performed by individuals who kept track of project details either in their heads or with lists of tasks to be performed. This approach worked well for simple projects; however, as the pace of the industrial revolution increased, it became obvious that this approach was not adequate for complex projects. It also became apparent that better methods were needed to anticipate potential project delays or problem areas so that additional resources and personnel could be diverted to ensure that the project was completed with a minimum of delay. One of the first innovations to aid in project management came in the early 1900's through the work-vs-time chart introduced by H. L. Gantt. This horizontal bar chart has since been named the Gantt chart, and gives a visual indication of the flow of progress for a given project. As an example, Figure 9–9 shows a Gantt chart for the operation of a car wash by a group of six college students. The chart gives a visual indication of the fact that certain activities cannot begin until others have been completed.

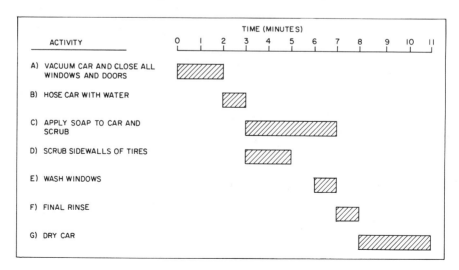

Figure 9–9 Sample Gantt chart for a carwash operated by six college students.

The disadvantage of the Gantt chart is that as projects become more and more complex, it gives no clear indication of how system dependencies can influence project progress, nor does it give any indication of how to manipulate the project resources to increase project efficiency. To satisfy these requirements a network method is needed.

The PERT–CPM network method

For a number of years the Gantt chart was the best project scheduling method available; however, in the mid 1950's several groups simultaneously began developing managerial approaches based on network analysis.

PERT (Program Evaluation and Review Technique) was developed in 1957 to aid in the planning, scheduling and control of the U.S. Navy's Polaris missile project. This project involved over 3000 different contracting agencies working on the program. Prior to the development of PERT the Navy had experienced substantial delays on all of its contract projects. As a result of the PERT method, the Polaris project was completed ahead of schedule. The PERT method is now one of the established tools for project management.

At about the same time as PERT was being developed by the Navy, the E. I. du Pont Co. developed the Critical Path Method (CPM) to meet its project management needs. The two methods are surprisingly similar. The major difference in their approach lies primarily in the manner in which activity times are established. In PERT three time estimates are established for each activity whereas for CPM only a single time estimate is used. Aside from this difference, the methods are quite similar and are usually treated in the literature as a single method known as PERT–CPM.

The basic elements in a PERT–CPM network are "activities" and "events." Activities are the operations in the project and, as such, consume time and resources. Events represent the beginning or end of an activity and occur as a milestone at a specific point in time. In a PERT–CPM network, activities are represented graphically by arrows and events are represented by circles, as shown in Figure 9–10. For ease of identification, events will be designated by number

Figure 9–10 The key elements in a PERT-CPM diagram are activities and events.

labels and activities will be designated by letter labels. To construct a PERT–CPM network diagram, the manager first makes a list of all tasks that must be completed in order to finish a project. Next he or she should consider what activities must be completed before other tasks can be started. By considering what milestone events will signal the completion of a given task the manager can construct a diagram such as the one shown in Figure 9–11. As shown in this figure, normally expected times for activities are indicated above or below the activity labels. Thus in Figure 9–11, activity A takes 2 units of time, activity B takes 2 units of time and so on. In this diagram it becomes obvious that activity E cannot start until both activity C and D have been completed.

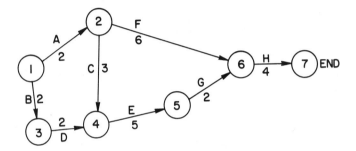

Figure 9–11 Example PERT-CPM netwook diagram.

In using PERT–CPM network diagrams three parameters are of importance. These are:

(a) The earliest expected event time (T_e)
(b) The latest allowable event time (T_L)
(c) The critical path

To determine the earliest event time for each event, the starting event is given a time value of $T_e = 0$, and subsequent earliest event times are computed by adding the values for the preceding activities. When more than one value is possible, as is the case for event 3, the largest T_e value is used. This type of event is called a merge event. Thus the earliest event times for the diagram of Figure 9–11 are indicated in Figure 9–12.

To determine the latest event time T_L for each event, the last event of the network is assigned the value $T_L = T_e$. The procedure then is to work backwards through the diagram, subtracting activity times of intervening activities. When

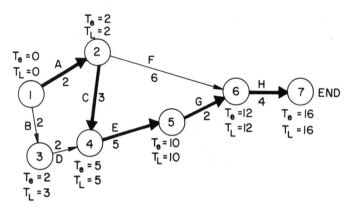

Figure 9–12 PERT-CPM diagram of Figure 9–11 showing earliest expected event times, latest allowable event times, and the critical path.

more than one value for T_L is possible, as is the case for event 2, the value used is the smallest value.

The critical path is the longest path through the system diagram and thus gives an indication of the minimum time in which the project can be completed. The critical path can be traced by identifying the path along which $T_e = T_L$. This path has been shown with a bold line in Figure 9–12. Although in this example only one critical path exists, it is possible to have more than one critical path. The total project times along these critical paths would, of course, be the same. The significance of the critical path is immediately obvious. If one shortens the activity times along the critical path by the addition of resources and manpower, one can speed up the overall project completion time. On the other hand, activities not on the critical path are said to have "slack time." These paths will not slow the overall project if they are allowed to take longer to complete by an amount less than or equal to the slack time.

In the overall operation of a project, the PERT–CPM diagram should be updated periodically in order to account for differences between expected activity times and actual activity times. This procedure allows the project manager to determine if the technique of diverting manpower and resources to the critical path has been effective. Since complex projects can have PERT–CPM diagrams that are rather involved, computer programs are now available to assist the project manager in identifying the critical path (Figure 9–13).

9.5 Summary

Organization and planning can often spell the difference between a successful project and a total failure. This is especially true for the design of large, complex

Figure 9–13 Engineers review data and analysis reports to provide up-to-date information for project administration. (*Courtesy* Bell Laboratories.)

systems in which there are extensive interactions among the elements. For this reason engineers should learn to use the systems design approach and the PERT–CPM method. As the technological pace of our world continues to accelerate, these topics will continue to become more important to engineers and engineering technologists.

PROBLEMS **9.1** Give three examples of systems that would best be treated by the traditional design method rather than by the systems design approach.

9.2 State an example of a system in which optimization of a subsystem can lead to a poor solution for the overall system.

9.3 Using the college catalog and the program of study you have chosen, draw a PERT/CPM network for your plan of study. Keep in mind that many engineering courses have prerequisite requirements. If a student could take an unlimited number of courses at any time, what is the minimum time required to complete the program?

9.4 Why is it important for an engineer to recognize the critical path in an engineering project?

9.5 For the Gantt chart shown in Figure 9–9, draw a PERT/CPM diagram. If more manpower were available to perform this project, where would it best be applied?

9.6 Classify the following projects as either systems design or traditional design:

(a) Design of the NASA Space Shuttle

(b) Design of a new type of coat hanger

(c) Design of a new mass transit system for a large city

(d) Design of an automobile

(e) Design of a new system to fasten papers together

(f) Design of a new type of desk lamp.

9.7 Make a PERT/CPM diagram that depicts the steps used in the following processes. (Estimate the times required where necessary.)

(a) Changing the oil in an automobile

(b) Brushing your teeth

(c) Baking a cake

(d) Designing a footbridge over a small stream

(e) Registering for courses at your college or university.

9.8 For the diagrams in problem 9.8 find the critical path.

REFERENCES

AGUILAR, R. J., *Systems Analysis and Design*. Englewood Cliffs, N.J.: Prentice-Hall, Inc., 1973.

ELMAGHRABY, S. E., *Activity Networks*. New York: John Wiley & Sons, 1977.

FROSCH, R. A., "A New Look at Systems Engineering." *IEEE Spectrum,* vol. 6, no. 9, 1969, p. 24.

"MEGASTAR" (The Meaning of Energy Growth: An Assessment of Systems, Technologies, and Requirements) with the Participants and Staff of the Auburn–MSFC Engineering Systems Design Summer Faculty Fellowship Program, NASA Contract NGT-01-003-044, October 1974.

MOORE, L. J., and E. R. CLAYTON, *GERT Modeling and Simulation*. New York: Mason/Charter Publishers, Inc., 1976.

THOME, P. G., "The Systems Approach—A Unified Concept of Planning," *Aerospace Management*. General Electric Company, Missile and Space Division, vol. 1, no. 3, Fall/Winter, 1966.

VACHON, R. I., et al., "A Training Exercise in Systems Engineering Design," *Engineering Education,* vol. 60, no. 8, 1970, pp. 819–822.

WAGNER, H. M., *Principles of Operations Research.* Englewood Cliffs, N.J.: Prentice-Hall, Inc., 1969.

WHITEHOUSE, G. E., *Systems Analysis and Design Using Network Techniques,* Englewood Cliffs, N.J.: Prentice-Hall, Inc., 1973.

Tilt-rotor research aircraft. (*Courtesy* NASA.)

Design projects

10

Every engineer needs to develop skill in applying the design process. Like other professional skills, engineering design requires knowledge, experience, and practice in order to perfect and to maintain competence. The best way to learn the professional skills associated with design is to experience this engineering process firsthand. This chapter provides a number of practical practice design experiences for engineering training.

Selecting or creating design projects at the introductory level in an engineering curriculum is like walking a tightrope. It is easy to oversimplify or overspecify a project, and the project then becomes too simple. On the other hand it is easy to formulate design projects which are well beyond the technical capabilities of undergraduate students. Yet, since projects have so many advantages, it is well worth the effort to try to walk this tightrope.

The advantages gained from a design project include the experience in working as a team member, the motivation from working on a real, tangible problem, and the integration of the successive phases of the design process into a single continuous sequence. A design project also allows the young engineer an opportunity to exercise his or her talents at creating and communicating ideas.

This chapter contains example projects which have successfully been used in classrooms at the undergraduate level. The projects are organized in a specific way in order to provide a clear, uniform presentation of their requirements. Because of variations among schools and students, minor adjustments may be made in these projects to adapt them to a particular classroom situation. For example, the instructor may ask for the utilization of specific skills from previous courses in the analysis of your design ideas.

The projects have been selected because they provide maximum opportunity for experiential learning in a way that is both interesting and challenging. For the young engineering professional who is just entering the design project environment, an exciting domain awaits.

10.1 Project 1: A water lens

Project statement

The purpose of this project is to design a low-cost lens which could be used in collecting solar energy. Details of the theoretical design will be developed graphically, and experimental testing will be used to verify these predictions. A use for this design also will be postulated as part of the project output.

Background information and suggested approach

The optics involved in this project are, briefly, as follows. A ray of light will be bent when entering and again when leaving the water. The relationship which expresses this is shown in Figure 10–1.

The value of the index of refraction between water and air is a ratio of the velocities of light in air and in water. Light travels 1.33 times faster in air than in water. Lens systems such as those shown in Figure 10–2 are an application of this physical property. In this project the light will also pass through a thin film of plastic which is used to form the lens. However, if the plastic film is uniformly thick, the direction of the light ray will not be changed; it will be displaced slightly, however, as shown in Figure 10–3. In both views of Figure 10-3, the dashed line is the direction of the ray if the plastic film is ignored, and the solid line is the actual direction of the exiting light ray. The thin plastic films are approximately 0.1 mm thick, so their contribution to the displacement of the light ray is very small and may be ignored.

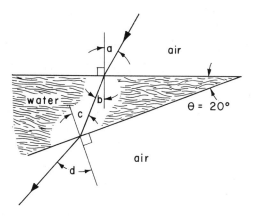

Figure 10–1 Refraction of light rays.

Note: $c = b + \theta$

$$\frac{\sin a}{\sin b} = \frac{\sin 30°}{\sin 22°} = \frac{\sin d}{\sin c} = \frac{\sin 63°}{\sin 42°} = 1.333$$

Figure 10–2 Plastic lenses focusing sunlight on solar cells. (*Courtesy* Solar Energy Research Institute—Sandia Laboratories.)

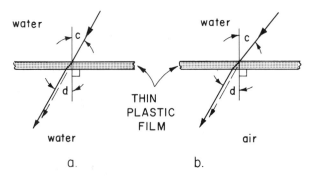

Figure 10–3 Light passing through plastic film.

A lens made from a sagging plastic film containing water will have a horizontal plane for a top surface and an unknown curve for a bottom surface, as shown in Figure 10–4. The plastic film could be taped to the outside wall of the container and could be adjusted to differing amounts of slack to obtain a variety of curves. The container could be circular (cylindrical) which would focus approximately to a point (or to a circular area), or the container could have parallel edges and form a trough of water which would focus to a line (or a rectangular band). In either case, if the container is level, the curve formed by the plastic film would be symmetrical about a center line. At a typical point P, shown in Figure 10–4, the film can be assumed to be sloping at an angle θ which corresponds to angle θ in Figure 10–1. Obviously θ varies across the film.

In order to predict the focus of your lens, you are to build a model with approximately 30 mm depth of water using a container of your choice. The container should have a partially open bottom, as later on we will experimentally check the focus of your lens. You may use heavy paper or cardboard to build the container. Scissors, water, and clear plastic film will be required for these tests.

After filling your lens with water, you must next devise an accurate method of measuring depths from above or measuring heights from below. Your choice of measuring may depend on the shape of your container. Use these measurements to make a full-scale drawing of your lens. The curve should be smooth and accurate, as it is the basis for your graphical determination of the focus of your lens.

After drawing the curve, carefully select a group of parallel rays ranging

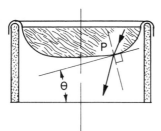

Figure 10–4 Light exiting from plastic and water lens at point P. The tangent to curve as P has a slope of Ø degrees.

from the center to the outer edge, and see what the focusing properties will be. Graphically construct the path for each ray using the calculation shown in Figure 10–1. Use an angle approximately equal to the sun's rays during the time of day you will be testing your model in the laboratory. Try to find the best area of focus, and measure its height from the lens. After this analysis, test your lens. This should be done with your instructor present. If the sun is available, use sunlight. If not, an electric light may be used.

Required output

Your project output should be a written report presenting your findings in an organized, logical fashion. Teams of students may work on each project. Be sure to include all names of the project team members on the title page of your report. Label all drawings, and give as part of your results the agreement between your prediction and the lens' actual focus. Make suggestions as to how the lens might be used and possible improvements in the design. The quality of the output of your project will depend particularly on how clear and easy to read your report is. Some possible questions to consider in your design study are:

1. How does focal-point location move with the change in angle-of-incidence of the sun's rays?
2. Can your lens be tilted? How far? In what direction can it be tilted and what effect does this have on the focus?
3. Water may evaporate. What effect would more/less water have on your lens? How much does your lens weigh? Would glass be better?
4. What effect does changing the slack or tension have on your plastic lens? If you allow more sag, does this improve the focus?
5. What other pertinent ideas occur to you? Can you see any unique shapes, materials, or uses for this general type of lens?

The project plan

In order to inform your instructor of your progress during the project, the following checkpoints are recommended. When you complete one of these steps, check with your instructor to review your progress before proceeding with the project:

1. Select the type, shape, and size of your model.
2. Build a sample model of your proposed design.
3. Make the graphical and experimental measurements of your design concept.
4. Organize the information you have obtained; analyze your results; retest if necessary.
5. Prepare your final report.

10.2   Project 2: Industrial robot

Project statement

The purpose of this project is to obtain design specifications for a robot. The robot will perform a specific task in an industrial plant. The task consists of picking up a small object at one location, moving it over to another location, releasing the object with a particular velocity, and simultaneously avoiding any interference with other stationary objects in the work area.

Background information and sample solutions

Figure 10–5 shows two views (a plan or top view and a front or elevation view) of a hazardous environment in an industrial plant. A small part comes out of a furnace at A and must be picked up and placed on a moving belt at B. The temperature at the exit of the furnace (A) is 150° C, and poisonous fumes occasionally are given off in this area. A nearby machine emits very loud noises, and the part being handled is potentially explosive. The task of transporting the part from A to B seems ideally suited for an industrial robot.

Figure 10–6 shows the robot which is to be programmed for this job. The robot has a hand which can open and close to grasp the part. The hand is attached to an arm which can rotate 270° and extend or contract, varying its length (L) from 600 mm to 1100 mm. The maximum speeds for the arm are 20°/s for rotation and 60 mm/s for extension (20°/s = .35 rad/s).

As a design engineer, you are asked to perform the calculations for the operation of this robot. This will require some drawings, some graphs, some mathematics, and possibly some computer programming.

A location for the robot must first be selected. Once this selection is made, measure the angle of rotation from A to B and the lengths of the arm at the initial position (L_o) and the final release position (L_f). Make a layout drawing to a scale of 1:10 showing the location of the robot, the angles, and the arm lengths required. Be sure these agree with the specifications for the robot which were previously given. This layout should include the features shown in Figures 10–5 and 10–6.

The second phase of the project is to prepare a graph of the rotation of the arm with the angular velocity as a function of time. The belt speed is 100 mm/s, and the hand on the end of the robot's arm should be traveling at the same velocity as the belt when the part is released. (Both magnitude and direction should be identical.) The speed of the end of the arm is equal to the radius times the angular velocity, $V = r\omega$.

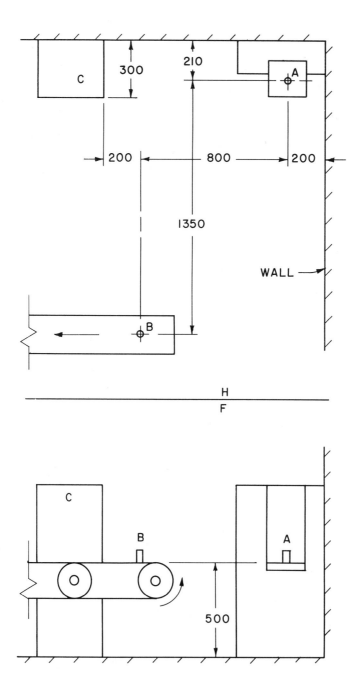

Figure 10–5 Plan and elevation views of work area. All dimensions in millimeters.

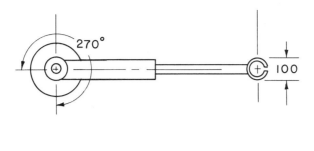

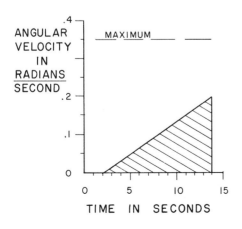

Figure 10–6   Top and front views of the robot.

A simple example is shown in Figure 10–7. Note that two seconds are re-quired at A for the arm to grasp the part and that the initial velocity is zero. The straight line curve indicates that the velocity begins at zero and slowly increases over the time interval to the required maximum. This motion is known as constant acceleration and requires the minimum amount of force to operate the robot. The area under this curve (a triangle) is the amount of rotation which has taken place. In this example, the area is:

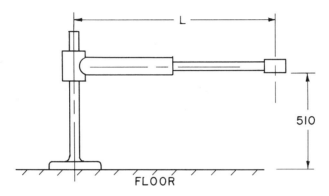

Figure 10–7   Rotational velocity of robot's arm.

$$\theta \text{ (radians)} = \frac{1}{2} \text{ bh} = \frac{1}{2} (14 - 2 \text{ s})(.2 \text{ rad/s}) = 1.2 \text{ radians}$$

$$1 \text{ radian} = 180°/\pi = 57.3°$$

$$\theta \text{ (degrees)} = 68.7°$$

If the arm is assumed to be 800 mm in length for this sample calculation, then the speed of the hand after 14 seconds is $V = r\omega = 800$ mm (.2 rad/s) = 160 mm/s.

This calculation and graph would be satisfactory if the arm needed to rotate 68.7° from A to B, and if the velocity of the end of the arm was 100 mm/s in the direction of the belt's velocity. If the angle, the speed, or the direction is not correct, the graph must be altered and/or the location of the robot must be changed.

In order to perform the task set for it, the robot's hand must not collide with the wall or other obstacles, such as the other machine at C in Figure 10–5. One means of avoiding a collision is to have the arm contract in length while rotating, and later extend after passing the obstacle. The hand is a 100-mm diameter cylinder, as shown in Figure 10–6, so a clearance of 50 mm (equal to the radius) is necessary between the center of the hand and any obstacle.

Figure 10–8 shows an example of an arm retraction and extension. Beginning at time = 2 s, the arm is retracted and continues until t = 6 s. In this example, retraction is negative area and extension is positive. At 6 seconds, the area under the curve is a total of – 60 mm. At t = 6 s, an extension begins and continues for 3 seconds until the arm is extended back to its original length at t = 9 s.

A graphical check of the position of the hand can be made combining the information contained in the graphs in Figure 10–7 and 10–8. The arm's angular position and length are drawn to scale for selected points in time. For example, if the arm began the cycle at a length of 800 mm, this motion would be as shown in Figure 10–9.

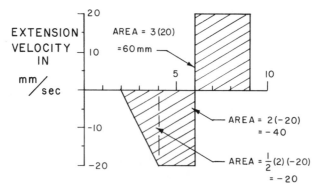

Figure 10–8   Velocity of robot's arm extension.

The calculations for the first two positions are as follows:

1. $t = 4$ s; P_4
   $\theta_4 = \frac{1}{2} (4-2) (0.1/3) = .033$ radians
   $\theta_4 = 1.9°$

   $L_4 = \frac{1}{2}$ bh (from Fig. 10–8)
   $L_4 = \frac{1}{2} (4-2) (-20) = -20$ mm
   length $= 800 - 20 = 780$ mm

2. $t = 6$ s, P_6
   $\theta_6 = \frac{1}{2}$ bh $= \frac{1}{2} (6-2) (2/3) (0.1)$
   $\theta_6 = .1333$ rad $= 7.6°$

   $L_6 = (-20) + $ bh $= -20 + (2) (-20)$
   $L_6 = -20 - 40 = -60$ mm
   length $= 800 - 60 = 740$ mm

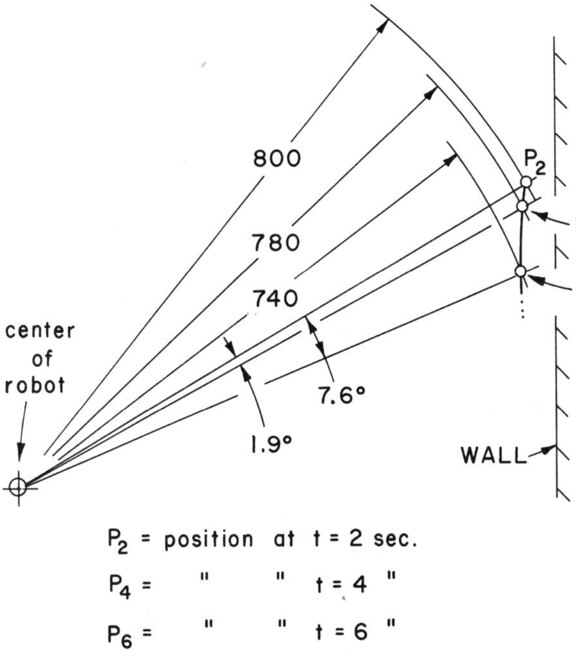

P_2 = position at t = 2 sec.

P_4 =   "   "   t = 4   "

P_6 =   "   "   t = 6   "

Figure 10–9 Plotting points on the path of the robot's hand.

If the direction of the hand is to be parallel to the belt at the time of release, the arm must be perpendicular to the belt at this instant. This assumes that the arm is not extending or contracting at the time of release, as shown in Figure 10–10a. If the arm is 800 mm in length and is rotating at 0.1 rad/s, then the velocity of the hand is $V = r\omega = 800 (0.1) = 80$ mm/s.

If the arm is simultaneously rotating and extending, then the direction is altered. If, for example, the arm is extending at 10 mm/s at the same instant it is rotating at 0.1 rad/s and its length is 800 mm, the magnitude and direction of the hand's velocity would be as shown in Figure 10–10b. $V_R = 80$ mm/s as in Figure 10–10a, and $V_E = 10$ mm/s, so the total velocity is $V = \sqrt{10^2 + 80^2} = \sqrt{6500} = $

80.62 mm/s in magnitude, and at an angle of $90° + \arctan (\frac{1}{8}) = 97.1°$ with the arm.

An industrial robot with many of the prescribed design characteristics is shown in Figure 10–11.

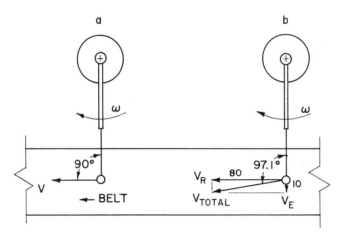

Figure 10–10 Velocity at release: (a) with no extension at release, (b) with extension at release.

Required output

Your project report should include the following graphs, all for the period of time from $t = 0$ at A until released at B:

1. Angular velocity of arm as a function of time, $\omega = f(t)$, similar to Figure 10–7.
2. Angular displacement (distance rotated) as a function of time, $\theta = f(t)$. This curve can be combined with that in number 1 above if the two scales are clearly shown. Angular displacement is the area under angular velocity.
3. Extension velocity as a function of time, $V_E = f(t)$, similar to Figure 10–8.
4. Length of arm as a function of time. This is the initial length plus the area under the graph in number 3 above and may be combined with that graph if you wish.

The project report should also include a drawing showing exactly where the robot is located, a complete plot of the path of the hand (similar to Figure 10–9), and the total time for the move from A to B.

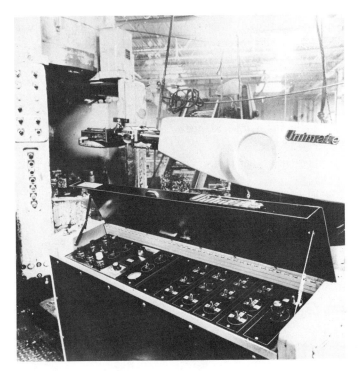

Figure 10–11 An industrial robot at work in the hazardous environment of a furnace. (*Courtesy* Unimation, Inc.)

This project report could include the following additional items:

1. Calculations and results for a complete cycle, showing the slowdown, stop, and return of the arm back to A. Additional time would be added to the four required graphs and to the plot of the robot's hand.

2. Add two seconds additional time onto the cycle from A to B. Assume that the hand takes two seconds to completely open and release the part. During this two second period following the arrival at B, the hand should move in a straight line path with the total velocity, *V*, constant at 100 mm/s. In order to accomplish this, the angular velocity must decrease slightly, and the arm must be extended slightly.

The project plan

In order to inform your instructor of your progress during the project, the following checkpoints are recommended. When you complete one of these steps, check with your instructor to review your progress before proceeding with the project:

1. A preliminary (trial) location, with calculations for velocity at B, and for the total angle which agrees with the location.

2. Trial extension with a solution for all four graphs.

3. Trial plot of the hand path.

4. Final path plot after any adjustments to the graphs.

5. Any additional items assigned.

Evaluation of your project should include the following criteria:

(a) Is your solution correct?

(b) Is your solution presented in a clear, logical manner?

(c) What have you added beyond the minimum requirements?

(d) Is your project neat and accurate?

10.3   Project 3: Folding seat

Project statement

Many aircraft are designed for special purposes, such as the Skycrane shown in Figure 10–12. The purpose of this project is to design a seat for a dual-purpose airplane. The airplane must be readily convertible from a freight carrier

Figure 10–12   Sikorsky S-64 Skycrane helicopter. (*Courtesy* NASA.)

to a passenger plane. The back of the seat must form a horizontal floor when the seat is in the collapsed, or freight-carrying, position. This floor should be continuous, with no cracks or openings between the seat backs, except possibly for recessed handles to be used to pull the seats upright.

Design information

The chair backs and seats are to be made of 3-mm–thick aluminum sheet metal. For comfort, the seats and backs may be covered with 50-mm–thick foam cushions. You do not need to consider the upholstering, but are to design the frame size and shape. The simplicity of the collapsing/setting-up operation is important. Certainly, pulling a lever or inserting a pin would be far less time-consuming than having to put on and take off many nuts and bolts for every seat. Armrests and headrests are not necessary, but may be included if you wish. The seat should not protrude more than 300 mm above the floor when collapsed and should be lower if possible in order to provide more cargo space. To obtain the dimensions of the passengers, refer to the anthropometric data in Appendix F.

Figure 10–13 shows some extruded aluminum shapes which may be used. Aluminum weighs 2.7 g/cm³, and the 3-mm–thick sheets weigh 8.4 kg/m². Simple pin joints, as shown in Figure 10–14, should be used. You may build a scale model of thin cardboard and soda straws if you wish.

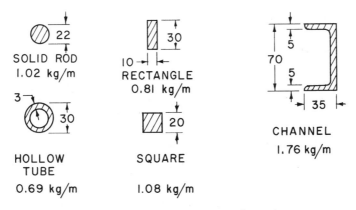

Figure 10–13 Extruded aluminum shapes; dimensions in millimeters.

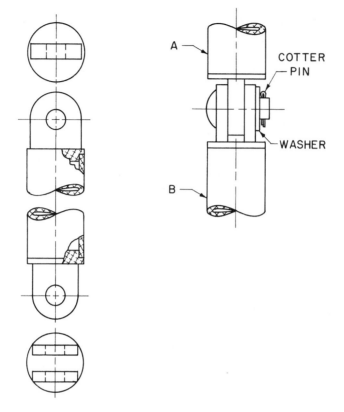

Figure 10–14 Typical pin joint joining parts A and B.

Required output

1. A scale drawing showing the front, top, and side views of the seat in both the upright and the collapsed positions. A few of the vital dimensions may be included, such as the overall width of the chair.

2. A list of parts required for each chair assembly.

3. A tabulated weight total. Try to estimate the weight to the nearest kg.

4. A pictorial view showing the steps necessary to convert the chair from one position to the other, with clearly written, easy-to-understand instructions.

10.4   Project 4: Crash escape compartment

Project statement

The purpose of this project is to'design the shape and surface of a compartment which will contain a single human being in a sitting position. This compartment will be ejected from a moving vehicle in event of a crash.

Background information and suggested approach

The compartment should consist of a small front end and a large rear end, with a transition section as shown in Figure 10–15. The upper part of the transition section should be transparent for visibility. The cross-sectional shapes of the compartment may be circular or oval as in Figure 10–16, elliptical as in Figure 10–17, or a combination of these as in Figure 10–18.

Other possible transitions, not shown, are circle to circle, (concentric), circle to circle (not concentric), ellipse to circle, ellipse to oval, oval to circle, oval to ellipse, oval to oval, etc. All can be either concentric (symmetrical) or non-concentric (non-symmetrical). Note that some combinations produce surfaces which have flat portions (planes), and some are entirely curved. For example, Figure 10–16 has a trapezoidal plane on each side, while Figures 10–17 and 10–18 are curved.

The wall thickness and the size of the passenger must be considered in the design of the compartment shell. The dimensions of human bodies are given in

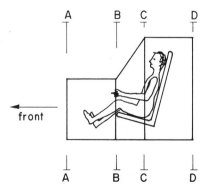

Figure 10–15 Typical compartment showing the three sections: front A-B, middle B-C, and rear C-D.

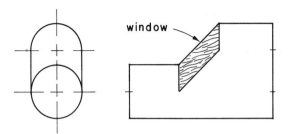

Figure 10–16 Compartment using circular and oval shapes.

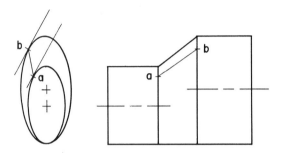

Figure 10–17 Compartment using elliptical shapes.

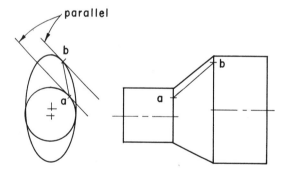

Figure 10–18 Compartment using circular and elliptical shapes.

Appendix F. The space available for wall thicknesses in the actual compartment are shown in Figure 10–19.

The transitional section (BC in Figure 10–15) may be slightly more complex to develop than the front (AB) or the rear (CD) section. A convolute surface can be used as a transition between any two curves. In your design, you will have two curves in parallel planes. If you mount these two curves perpendicular to a

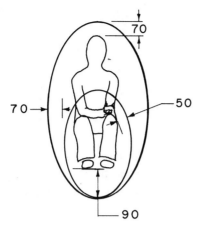

Figure 10–19   Minimum air space between passenger and compartment walls.

shaft and slowly roll these two curves over a plane (a piece of paper), the tracks the curves leave will be an outline of the development of the desired surface (see Figure 10–20). Since the front and rear sections will be some form of cylinders, the following information may be useful:

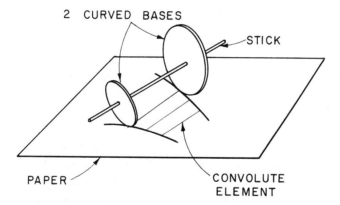

Figure 10–20   Constructing convolute elements for developing a transition piece.

Volume of cylinder (right, any shape base) $V = Ah$, where $A =$ area of base. If circular, $V = \pi r^2 h$ as $A = \pi r^2$.

Volume of cone $= (\frac{1}{3})\pi r^2 h$ for entire circular cone. If truncated, $V = (\frac{1}{3})\pi (r_1{}^2 + r_1 r_2 + r_2{}^2)h$ where r_1 and r_2 are the radii of the upper and lower bases.

Area of an ellipse $= \pi ab$

$$\text{Circumference of an ellipse} = 2\pi\sqrt{\frac{a^2 + b^2}{2}}$$

where: a = semimajor axis
b = semiminor axis

A planimeter may be used to measure the area of any unknown curve. See your instructor for aid in using this instrument. If an unknown shape is plotted on graph paper, the area may be approximated by graphical integration methods.

Required output

This project requires the construction of a 1:10 scale model of the compartment, as well as a written report. The model's skin can be thin paper, with stiff paper or cardboard used for the cross-sectional shapes. The report should contain:

1. Cross sectional areas at A, D (of Figure 10–15).
2. The length of each section A–B, B–C, and C–D.
3. The total surface area, including both ends.
4. The total volume.
5. A development layout to scale for each separate piece.

The project plan

The first consideration is to select a size and shape for the compartment which will provide adequate space for the passenger. After the sizes and shapes have been selected, each surface must be developed and measured. The model can then be constructed and will serve as a check on the accuracy of the development. Finally, the report must be prepared.

10.5 Project 5: Solar cooker

Project statement

Recent energy shortages have focused attention on alternative sources of energy, such as solar energy. This project involves the design, analysis, building, and testing of a low-cost solar cooker. The cooker will be used to cook a hot dog, using only the sun's rays as its source of energy. It will be built from corrugated cardboard and will use aluminum foil as the reflective surface.

Background information and suggested approach

An efficient design would require the use of an approximately parabolic reflector and would focus the rays into a heating zone (see Figure 10–21). If the cooker is designed to cook a single hot dog at a time, the rays could be focused into a line about 150 mm long. A true paraboloid of revolution focuses parallel rays into a point, but such a double-curved surface could not be developed from

Figure 10–21 A typical parabolic solar collector. (*Courtesy* Solar Energy Research Institute–Sandia Laboratories.)

cardboard. The parabolic surface may be approximated, however, by multiple zones (shown in Figure 10–22), each of which is a portion of a right circular cone. These hypothetical cones will all have the same axis but different vertices. The sunlight from a zone will be reflected or focused into a line segment along the axis, not a point. If the sizes of the zones are selected with care, this focus line can correspond to the length of the hot dog.

The shape of the parabola can be varied by using different values for the constant k in the general equation $y = kx^2$. The development of this type of figure

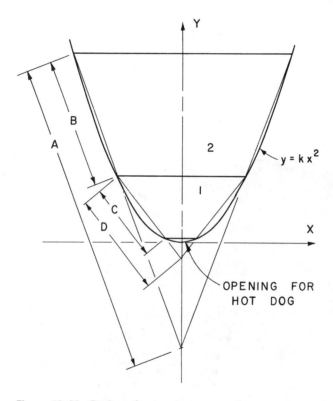

Figure 10–22 Design of a two-zone approximate paraboloid surface.

consists entirely of straight lines and circular arcs. The arcs too large for a compass may be drawn accurately enough with a pencil and a piece of string. The calculation of the angles used in the development is a simple proportion:

$$\theta = \frac{R}{L} 360°$$

where R is the radius of the circular base, and L is the length of the elements of right circular cone. These are shown in Figure 10–23. The development for the two zones shown in Figure 10–22 is shown in Figure 10–24.

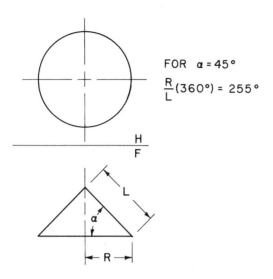

FOR $\alpha = 45°$

$\dfrac{R}{L}(360°) = 255°$

Figure 10–23 Dimensions of a right circular cone.

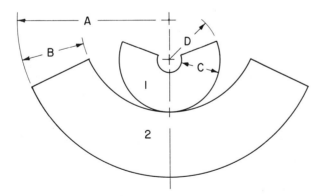

Figure 10–24 Development of the two truncated right circular cones shown in Figure 10–22.

Table 10–1 shows quantities needed in a sample calculation in both SI and English units for the estimation of the time required to cook the hot dog. Cooking the hot dog is assumed equivalent to boiling an equal weight of water.

Table 10–1.   Hot-Dog Cooking Information

Units	SI	English
diameter of opening of cooker	700 mm	27 in.
area of opening, πr^2	.385 m²	4 ft²
sun energy per minute approximately	$\dfrac{45\ 400\ \text{joules}}{\text{m}^2\ \text{min.}}$	$\dfrac{4\ \text{BTU}}{\text{ft}^2\ \text{min.}}$
energy into cooker	0.385 (45 400) = 17 500 J/min., (4.19 J will raise temp of 1 g water 1°C)	4 (4) = 16 BTU/min., (1 BTU will raise temp of 1 lb water 1°F)

The time required to cook is:

$$\text{(SI}\qquad 45\ \text{g}\ (80°\text{C}) \times \frac{4.19\text{J}}{\text{g}} \times \frac{\text{min}}{17500\text{J}} \times \frac{1}{.15} \cong 6\ \text{min.}$$

$$\text{(English)}\ 0.1\ \text{lb}\ (142°\text{F}) \times \frac{\text{min}}{16\ \text{BTU}} \times \frac{1}{.15} \cong 6\ \text{min.}$$

Actual cooking time was found to be eight to ten minutes, slightly more than the six minutes calculated.

Required output and project plan

The project will be undertaken in teams assigned by your instructor. Each team will need several large corrugated boxes, tape, and aluminum foil for lining the interior of the cooker. The hot dog may be impaled on a coat hanger or a meat thermometer if you want an instrumented experiment. In order to function correctly, the cooker must have its axis pointing toward the sun. This can be done visually by holding a meter stick in the center of the cooker in a position with no shadow.

The design project must include a drawing similar to Figure 10–25. In this drawing, the sun's rays enter the collector parallel to the axis and are reflected at angle θ which is equal to the angle of incidence θ. The length of the reflector zone, *A*, will determine the length of the cooking zone, or focal line. The scale drawing must be adjusted so that all zones focus onto the same line segment, and the location of this segment must be known. The hot dog must be located correctly to avoid uneven cooking.

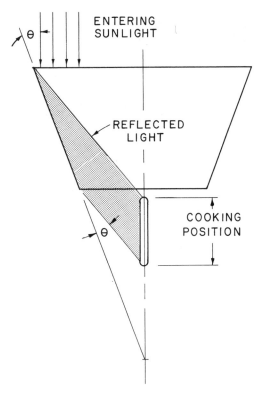

Figure 10–25   Focus zone of the cooker.

After this drawing has been approved by your instructor, begin the development of the cones. A scale layout of the development may help you minimize errors and material waste. All these drawings will be included in your written report. The cooker should be used on a sunny day, and the actual results compared to the theoretical results. The report should also include a discussion of the results and recommendations for improvement.

10.6   Project 6: An inflatable boat

Project statement

Inflated objects are used as structures to enclose tennis courts, as life rafts, as escape chutes (as shown in Figure 10–26), and in many other applications.

The purpose of this project is to design an inflatable boat capable of carrying a piano over water. A 1:10 scale model of the boat will be built and tested.

Figure 10–26 Inflatable escape chutes for ground emergencies on passenger aircraft. (*Courtesy* NASA.)

Background information and suggested approach

The piano is 600 mm deep, 1500 mm wide, 1200 mm high, weighs 500 kg, and must be placed in an upright position on the boat. Assume the loading and unloading will be done by an overhead crane.

When the piano is reduced to a 1:10 scale model, its dimensions are $\frac{1}{10}$ size, its area is $\frac{1}{100}$ and its volume and weight are reduced to $\frac{1}{1000}$. A block of wood weighing about 500 grams will be used to test your boat.

The first stage of the project is to decide on the geometric shape for the boat. Make sketches of several alternate shapes consisting of cylinders, cones, prisms, planes, etc. A set of scale drawings will then help determine the size of the boat.

An attempt should also be made to analyze the boat's stability. Since the piano must be transported in an inherently unstable upright position, your boat design should counteract this instability. Figure 10–27 shows the difference between a stable and an unstable floating object. When the force of weight, W, and bouyancy, F, are equal, the volume of water displaced is equal to the weight of the object. The force W acts through the center of gravity of the whole object, while the bouyancy force F acts through the center of the shaded area (the center

STABLE   UNSTABLE

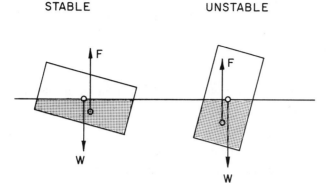

Figure 10–27 Stability analysis of a floating object.

of gravity of the displaced water). In the stable situation, the bouyancy force
tends to push the object back to a level or stable position. In the unstable state,
the object is going to tip over as *F* forces the object to fall to the right.

Assume the center of gravity of the piano is at the geometric center, where
the interior diagonals intersect. Since the raft is quite light in comparison to the
piano, assume this is also the center of gravity for the entire unit, boat and piano.
To analyze the system, assume an angle of tilt of 30°. Make a scale drawing of
your boat in this position, and see if the boat would tip over or if it would tend
to right itself. This same test may be made experimentally when your model is
complete.

Required output

1. A 1:10 scale model constructed using heavy paper, tape, scissors, and a
   waterproof spray. An alternate to a spray is to cover the bottom with a thin
   plastic film similar to that used to cover foods.
2. Scale drawings showing the size of the boat. Drawings should also be provided showing developments used in construction of the boat's surface. These
   drawings will provide information on the surface area.
3. Your preliminary ''idea'' sketches, showing alternative shapes which were
   considered.
4. Analysis with calculations of:
   (a) the surface area
   (b) the volume of air enclosed
   (c) the depth the boat will sink in the water, loaded and unloaded (its draft)

(d) the height of the boat's edge above the waterline (freeboard)

(e) the total area of material required, including waste, and what percent of material is waste

(f) the total length of all seams

(g) the cost; assume material is purchased in rectangular shapes at 1¢/cm², and seams cost 2¢/cm.

5. A pictorial view of model.

6. Results of testing the model in water.

7. Conclusions and recommended changes in design.

Project plan

This project may be done on an individual basis or as two-person teams. A team should provide results which are slightly more thorough than those of a project worked out by an individual alone.

10.7 Project 7: Ducts for connecting tanks

This project involves the design of two ducts to connect two supply tanks with a mixing tank. The project purposely does not include any textbook illustrations in order to give the student experience in interpreting word descriptions.

Background information and suggested approach

Three tanks with different geometrical shapes are to be connected by means of a system of welded straight ducts. The size and location of the tanks are as follows:

Tank A is a cube that is 6 m on each side. The base is horizontal, and one of the diagonals of the base is a point and the other is true length in the front view.

Tank B is a right rectangular prism that has a base 8 m by 15 m and a height of 6 m. The base appears true shape in the top view, and the 15 m side is true length in the front view. The lower left front corner of tank B is 5 m above, 10 m to the right, and 6 m in front of the lower right-most corner of the base of tank A.

Tank C is a right circular cylinder that has an 8 m diameter and a 6 m height. The base appears as a circle in the top view. The center of the base is located 5 m above, 5 m to the right, and 12 m in front of the bottom right-most corner of the base of tank A.

Tanks B and C contain fluids that will flow into tank A through a duct system using gravitational forces. The ducts must have a minimum internal cross-section of 8 m² and are to be fashioned from available surplus which consists of:

(a) one 12 m long duct with a square internal cross-section of 9 m²

(b) one 14 m long duct with a circular internal cross-section of 8 m²

(c) one 16 m long duct with an equilateral triangular internal cross-section of 8 m².

All of the above stock ducts have a wall thickness of 5 mm and are made of aluminum, which has a density of 2.7 g/cm³. The duct system intersects tanks B and C in a manner that allows complete drainage of the fluids. The straight portions of the duct system must have a slope of at least 10° to develop adequate fluid velocity towards tank A.

You are to design a duct system from the available stock that satisfies the preceding specifications.

Required output

The completed design project will consist of:

1. At least one freehand sketch of a possible duct configuration.
2. A complete scale drawing (1:250) of the duct system consisting of at least two orthographic views of the layout.
3. A tabulation of the slopes of each straight portion of the duct system.
4. A development of each section of the duct system (1:100).
5. A calculation of the total weight of the duct system and total length of all seams.
6. A model of the system (1:100).

Project plan

The project is an individual effort.

10.8   Project 8: Skeet trap

Project statement

This project involves the design of a clay pigeon launcher, commonly called a skeet trap. The skeet trap, shown in Figure 10–28, is a device which slings a disk-shaped projectile (clay pigeon) into the air by means of an arm and spring. This device has adjustments that allow the direction and angle of launch to be varied. The device is composed of a stand, spring, and launching arm. The projectile is launched by tripping a catch which releases a spring that rotates the arm and pigeon in a circular path until the arm contacts a stop. When the arm strikes the stop, the pigeon leaves the launcher with velocity V and a launch angle of θ. When it leaves the launcher, the pigeon is both rotating and translating. The rotation causes a gyroscopic stability that keeps the pigeon on a smooth path.

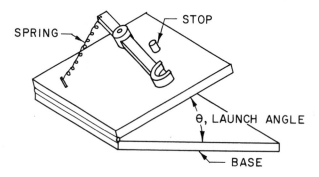

Figure 10–28   Typical skeet trap.

Background information and suggested approach

A number of relationships will effect the design of the skeet trap. First, it would be wise to determine the effect the angle θ (the launch angle in Figure 10–28) and the initial velocity have on the range and altitude of the pigeon. If we ignore aerodynamic effects of air on the pigeon and assume the terrain is level, these relationships are:

(a) Maximum range, $X = \dfrac{V^2}{g} \sin(2\theta)$

(b) Maximum altitude, $Y = \dfrac{V^2}{2g} \sin^2\theta$

where: V   = initial velocity, m/s
        θ   = launch angle
        g   = gravitational acceleration, 9.81 m/s²
       X,Y = distances, m.

    If we decide on an initial range of 50 m and a maximum height of 10 m, the two equations above can be solved for two unknowns, *V* and θ.

$$50 = \frac{V^2 \sin (2\theta)}{9.81} \tag{1}$$

$$10 = \frac{V^2 \sin^2\theta}{2(9.81)} \tag{2}$$

from Eq. (1), 9.81 (50) $\sin^2\theta = V^2 \sin (2\theta) \sin^2\theta$
from Eq. (2), 19.62 (10) $\sin^2\theta = V^2 \sin (2\theta) \sin^2\theta$
490.5 $\sin^2\theta = 196.2 \sin 2\theta = 196.2$ (2 sin θ cos θ)

$$\frac{\sin \theta}{\cos \theta} = \tan \theta = \frac{2(196.2)}{490} = .40$$

$$\theta = 21.8°$$

from Eq. (2), $V^2 = \dfrac{196.2}{\sin^2\theta} = \dfrac{196.2}{(.371)^2} = \dfrac{196.2}{.149} = 1318$

$$V = 36.3 \text{ m/s}$$

    Thus, if the initial velocity is 36.3 m/s, and the angle of the launch is 21.8°, the pigeon will land at a point 50 meters from the launch site and will have attained a maximum height of 10 meters in flight.

    A second relationship which affects the design of the skeet trap is the influence the spring constant has on the initial velocity. If we assume the kinetic energy of the arm at the time of release is equal to the potential energy of the elongated spring, these two quantities may be expressed as follows:

$$\text{kinetic energy} = \tfrac{1}{2} \, I\omega^2 \tag{3}$$

$$\text{potential energy} = \tfrac{1}{2} \, KX^2 \tag{4}$$

where: I = moment of inertia of the arm, including pigeon, kgm²
ω = angular velocity of arm at launch, rad/s
K = spring constant, N/m
X = distance the spring is stretched, m

The geometry of the spring stretch must be known in order to cock the trap to the desired position to obtain a particular velocity. For example, assume the user wants an initial velocity of 36.3 m/s in order to send the pigeon 50 m, as shown in the previous calculation. If the radius of the arm (center of pigeon to center of axis) is 0.5 m, then

$$V = r\omega \tag{5}$$

$$36.3\text{m/s} = (.5 \text{ m})\omega$$

$$\omega = 72.6 \text{ rad/s}$$

In order to solve Eq. (3) for kinetic energy, the moment of inertia of the arm must be found. The moment will be found in three separate parts, as shown in Figure 10–29. The two ends of the arm are each considered as slender rods with the axis at one end. The equation for the mass moment of inertia (L), for this shape is

$$I = 1/3 \text{ ML}^2 \tag{6}$$

where:

M = mass of rod, kg
L = length of rod, m.

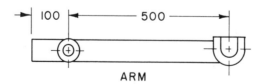

ARM

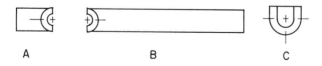

A   B   C

Figure 10–29 The arm divided into three parts.

If the arm material were 1.0 kg/m, the masses of A and B are:

$$M_A = 0.1 \text{ kg and } M_B = 0.5 \text{ kg.}$$

The moment of inertia for these two are:

$$I_A = (0.1) \ (.1)^2 = .001 \text{ kg m}^2$$

$$I_B = (0.5) \ (.5)^2 = .125 \text{ kg m}^2$$

The third part of the arm consists of the cup and the pigeon. For simplicity, assume the mass of the cup and the pigeon is located at the center of the pigeon, 0.5 meters away from the axis. The masses are:

pigeon: 0.12 kg

cup (assume 50 cm³ of aluminum @ 2.7 g/cm): 2.7 (50 = 135 g = .135 kg

total mass = 0.120 + 0.135 = 0.255 kg

The moment of inertia for a point mass is $I = ML^2$ where $1 =$ distance from axis. The moment of inertia for the cup is

$$I_C = (0.255) \ (.5)^2 = .0638 \text{ kg m}^2$$

The moment of inertia for the entire arm is the total.

$$I = I_A + I_B + I_C = 0.001 + 0.125 + .0638 = 0.190 \text{ kg m}^2$$

With this result, the kinetic energy can be computed, using Eq. (3)

$$\text{K.E.} = \tfrac{1}{2} \ I\omega^2 = \tfrac{1}{2} \ (0.190) \text{ kg m}^2 \times (72.6)^2 \frac{\text{rad}^2}{\text{s}^2}$$

$$\text{K.E.} = 500 \ \left(\frac{\text{kg m}}{\text{s}^2} \right) \text{ m} = 500 \text{ N m} = 500 \text{ J}$$

Since we also assume the potential energy to equal this amount, Eg. (4) can be written $500 = \tfrac{1}{2} \ K \ X^2$. Figure 10–30 shows the geometry involved. When the arm is cocked at a specific angle α, the spring is stretched a distance X, where X =

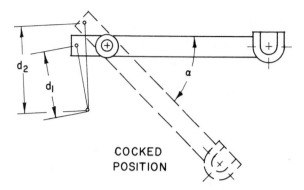

Figure 10–30 Geometry of arm and spring.

$d_2 - d_1$. The combination of this geometry, angle α, and distance X (all arbitrarily chosen) determines the spring constant K. Using Eg. (4), if X = 0.2 m,

$$500 = \tfrac{1}{2} \, (K) \, (0.2)^2 = .02 \, K$$
$$K = 25{,}000 \text{ N/m}.$$

Once a spring constant is calculated and a spring is chosen for the skeet trap, the trap could be calibrated so various heights and distances could be chosen by adjusting the launch angle and the angle at which the arm is cocked for launch.

The designer must also consider the amount of pull necessary to cock the arm. If the spring constant is 25,000 N/m and the spring is stretched 0.2 m, the force exerted by the spring is (0.2) (25000) = 5000 N. If the user pulls the cup back to cock the arm a distance of 0.5 m from the axis, and if the spring is attached to the arm a distance of 0.15 m from the axis, this is an advantage of 0.5/0.15 for the user. Thus, the user would have to exert a force of (0.15/0.5) (5000 N) = 1500 N to cock the arm. This force is too large and would render the design unsatisfactory. For the average person, the maximum force to be expected would be in the 400 N to 500 N range.

To decrease the force required to cock the arm, several factors can be altered. The geometry (such as the length of the arm and the position of the spring on the arm) affects the mechanical advantage and can easily be shifted. The spring constant selected and the distance the spring is stretched also can be changed. Some of these changes (as in the length of the arm) will also change the performance of the arm, its moment of inertia, its mass, etc. Thus, it will probably be necessary to go through several iterations to arrive at a satisfactory solution.

Some important design criteria are summarized below:

(a) A pigeon is 100 mm in diameter, 25 mm high, and 0.12 kg in weight.

(b) The range of the trap should be about 50 m, with available height ranging between 10 m and 15 m.

(c) The force required to cock should not exceed 500 N; 400 N would be better.

Required output and project plan

Your report should include:

1. A drawing of the arm, showing all important dimensions.
2. A drawing showing the geometry of the spring stretching and cocking force for a specific range.
3. Sample calculations demonstrating clearly that your design works.
4. A description of any extra features you wish to incorporate in your design.

10.9   Project 9: Solar mirrors

Project statement

The purpose of this project is to design a battery of mirrors which will reflect sunlight onto a boiler as a means of collecting solar energy (see Figure 10–31). The boiler will then furnish steam to generate electricity to heat buildings or for other useful purposes. A proposal was made to NSF to build such a system in Africa using a 1.14 km square field with about 14,000 mirrors, each mirror about 6 m^2 and each aimed at a spherical boiler on a tower 260 m high. The orientation of each mirror was to be controlled by a remote computer.

Background information and suggested approach

Your project is to design a similar type of installation. Choose your own land site, and show its topology and area as well as compass direction on drawing. Decide on the size, shape, spacing, and general layout of your mirrors, and the location of your boiler. The size and shape of your boiler may well, of course, be influenced by the size and location of the mirrors. For example, a larger mirror would need a larger target for its reflected sunlight.

The mirrors' movement will probably be controlled by two variables: the compass direction and the angle of tilt. These two angles are defined as follows:

γ: the azimuth angle; deviation from south, in degrees, with east positive and west negative. For mirrors facing south, $\gamma = 0°$.

Figure 10–31 A field of nearly 200 mirrors in the Department of Energy solar thermal power generation facility. (*Courtesy* Solar Energy Research Institute–Sandia Laboratories.)

s: the slope angle; and angle measured from a horizontal plane in degrees. For horizontal mirrors, $s = 0°$.

Figure 10–32 shows a mirror with an orientation of $\gamma = 37°$ and $s = 48°$.

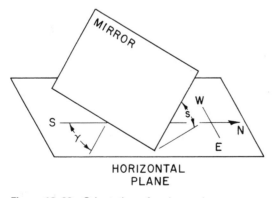

Figure 10–32 Orientation of a plane mirror.

Light is reflected from a plane mirror in a direction which makes an angle equal to the angle of incidence, as shown in Figure 10–33. In this figure, the incoming light is line AB, the reflected light is BC, and the normal to the plane is line BD.

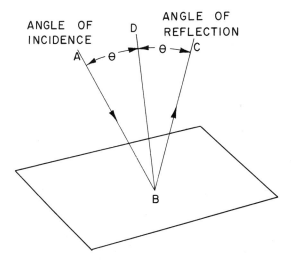

Figure 10–33 Angle of incidence and reflection.

The direction of the sun's rays can be determined approximately if three factors are known: (1) the latitude of the mirror's location, (2) the day of the year, and (3) the time of day. The equations for this calculation are

$$\delta = 23.45° \sin \left[360° \left(\frac{284 + N}{365} \right) \right] \qquad (1)$$

$$\cos \theta = \sin \delta \sin \phi + \cos \delta \cos \phi \cos \omega \qquad (2)$$

where: N = day of year, with January 1 = 1
   δ = angle of declination, in degrees
   θ = angle of incidence for a horizontal mirror (see Figure 10–33) in degrees
   ϕ = angle of latitude, 38° for Charlottesville, Va.
   ω = hour angle (noon = 0°; each hour = 15°)
      example: 11:00 am, $\omega = -15°$
      14:30 pm, $\omega = +37.5°$

If the position of the center of the mirror is known with respect to the center of the boiler, then both lines AB and BC are known. The only unknown is the orientation of the mirror. The normal to the mirror, line BD, is the bisector of the angle ABC, and lies in the plane ABC.

For example, assume a mirror is positioned with its center at point M, and a spherical boiler has its center at point P, as shown in Figure 10–34. Lines LM (sun's ray) and MP are both known. The unknown is line MN, the line perpen-

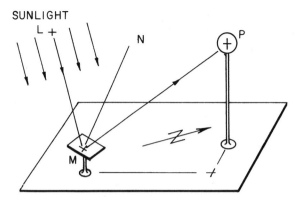

Figure 10–34 Relationship between sun, mirror, and boiler.

dicular to the plane mirror. To find MN graphically, the angle LMP is bisected. Since this angle can only be bisected in the view where the plane LMP is shown in true shape, this may require two auxiliary views (see Figure 10–35).

In this example, assume a mirror is put into position at 38° N latitude on April 1, 1982 at 1:30 pm.

From Eq. (1),

$$\delta = 23.45° \sin \left[360\left(\frac{284+91}{365}\right) \right] = 23.45° \sin (369.9°)$$

$$\delta = 23.45° \sin 9.86° = 23.45° (.1713) = 4.017°$$

From Eq. (2)

$$\cos \theta = (\sin 4.017°)(\sin 38°) + (\cos 4.017°)(\cos 38°)(\cos 22.5°)$$

$$\cos \theta = (.070052)(.615661) + (.997543)(.788011)(.923879)$$

$$= (.0431267) + .7262385$$

$$\cos \theta = .8693653$$

$$\theta = 39.7°$$

As shown in Figure 10–35, at 1:30 pm the sun's ray are approaching the mirror M from the southwest at (13:30–12:00) = 1.50 hours × 15° = 22.5° past south (see top view). The angle of incidence of 39.7° is seen in the front view. Draw a horizontal line from P (the boiler) which intersects LM at X. The first

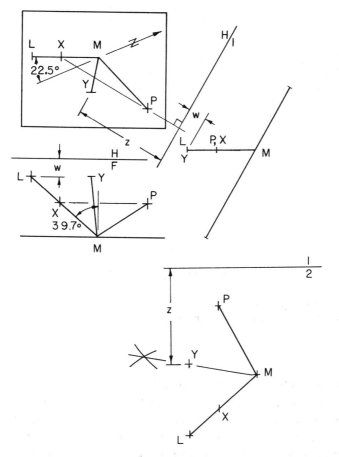

Figure 10–35 Graphical determination of mirror's position using two auxiliary views.

auxiliary view is drawn looking in a direction parallel to true length line PX in the top view. This auxiliary shows the plane LMP in edge view. The second auxiliary is drawn using reference line 1, 2 parallel to the edge view of LXMP. In this auxiliary view, point Y is determined by bisecting angle LMP. Point Y may be transferred back into views I, H, and F using projection rays and proper distances. For example, when point Y is found in the second auxiliary view, the distance Z can be measured and then used to locate point Y in the top view. Line MY, the line normal to the mirror, furnishes the information needed to position the mirror M such that the sunlight will be reflected to the solar boiler at P.

Figure 10–36 shows line MY in a top, front, and auxiliary view. The mirror should be tilted toward the southeast, in the direction of MY. The angle s which the mirror makes with the horizontal (ground) shows in the third auxiliary view, the view which shows MY in true length. The problem is now solved for a single mirror for a specific time, date, and location. The mirror M should be tilted at

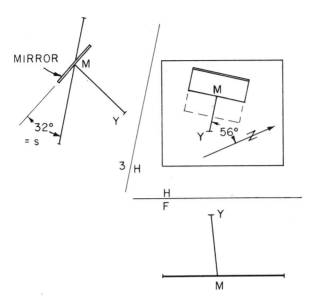

Figure 10–36 Using an auxiliary view to determine compass direction for mirror orientation.

32° toward S56°E in order for its sunlight to hit the boiler.

One question which now arises is how often the mirrors should be moved. In other words, what error will occur if the mirror is in the same position at 2:00 pm as it was at 1:30 pm? Also, the mirrors may shade each other when the sun is at a low angle in the early morning and late afternoon. This will depend on the terrain, the height of the mirrors, and their spacing. To determine the accuracy of your calculation, a scale model could be put outside to check your results against the actual sun's rays.

If you wish to become involved in computer programming, a program could be written to calculate compass direction and angle of tilt for a specific mirror at chosen time intervals. The program could be expanded to include mirrors at various specific locations with respect to the boiler.

Required output

This project lends itself to a number of different activities. Your instructor may want to assign some or all of them.

1. Prepare a scaled layout drawing and/or a pictorial drawing showing overall dimensions, typical spacing dimensions, and height of boiler.

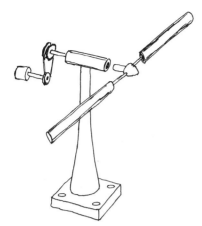

Figure 10–37 An idea sketch.

2. Prepare an idea sketch or drawing showing the size, shape, and type of motion of a typical mirror, similar to the idea sketch for a windmill in Figure 10–37.

3. Solve graphically for compass direction and angle of tilt of a specific mirror. For simplicity you may wish to use a mirror-boiler combination which lies on a N-S line and use noon for the time of day. The general case, as solved in the sample calculation, is more difficult.

4. Build a small model of mirror and boiler and test your calculations. Calculate for a time and date during classtime, and hope you get a sunny day.

5. Calculate how often your mirror should change orientation, or what error would occur in a specific time interval. If the boiler is quite large in comparison to the mirror, loss from the mirror would be reduced. Yet a large boiler will produce a bigger shadow, and the reflected sunlight will be spread over a larger area, reducing the temperature on the boiler's surface.

6. Assume incoming sunlight at 45 500 J/m²min, and assume a 10% efficiency (which means that 90% of the energy is lost and 10% is used to heat the water). If 4.186 J raise the temperature of 1 g of water 1°C, and if entering water is 15°C, then 4.19 (100° − 15°) = 356 J is required to bring 1 g of water to 100°C. To change 1 g of water, at atmospheric pressure, to steam requires an additional 2084 J. Thus, 2440 J are needed to change each gram of entering water to steam. At what rate should water enter the boiler?

7. Try to determine for a specific day of the year, what time in the morning and in the evening the system would receive only 50% of the maximum midday energy due to the mirrors shading each other. If the system is shut down, at what position should the mirror be placed?

10.10   Project 10: Stair-climbing wheelchair

Project statement

The purpose of this project is to design a powered wheelchair or similar vehicle which would go up and down stairs in most buildings.

Background information and suggested approach

A design which is complete in detail is not required, but some ideas, concepts, and overall dimensions should be included in your report. Tinkertoys may be used to build models if you wish. If you build a model for experimentation, be sure to specify a scale. The size of the power pack should be approximately 30 000 cm³. Typical human dimensions are available in Appendix F of this book. If possible, inspect a standard, non-climbing wheelchair. Measure the dimensional characteristics of public buildings, such as the size of staircases, doorknobs, elevators, hallways, etc. Can the user reach and open a door so the wheelchair can pass through? Figure 10–38 shows a powered wheelchair which cannot climb stairs.

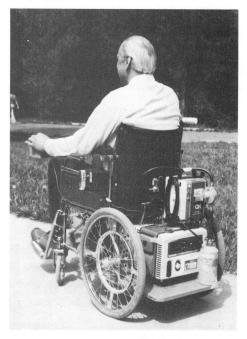

Figure 10–38   Power wheelchair. (*Courtesy* Mechanical and Aerospace Engineering Department, University of Virginia.)

Required output

The main objective of this project is to provide an opportunity for creative design and to provide an experience in expressing your own ideas. The ideas should be expressed clearly through drawings, sketches, and words, in the form of an illustrated report. The report should provide a description of the actual device, explaining the various features and how they work. Some dimensional specifications for size may be given. For example, the minimum width of a doorway which will allow the vehicle to pass should be noted. If the chair will go up some stairs but not others, describe the limitations. To keep the wheelchair from tipping over, you might consider having the operator in a different position when going up from going down stairs. If some manipulations by the operator are necessary to engage the stair-climbing feature, describe how the wheelchair changes to and from a climbing wheelchair.

Point out any unusual features your design incorporates, tell why they are included and what the advantages and disadvantages of your design are. Include any information you think a potential buyer would want to know. If you wish, you may try to estimate weight, cost, etc., but this is not required and may be difficult to determine exactly.

Your team may wish to build a model in order to test design ideas, and you may use a model in your 15-minute final presentation during the final class period; however, a model is not required. If your design has moving parts, show clearly how they move (e.g., if they rotate about a vertical axis) and how far they can move (e.g., this leg can extend from 500 mm to 1000 mm).

10.11   Project 11: Fixed mirror solar collector

Project statement

A variety of space geometry systems are used to make solar collectors more efficient(see Figure 10–39). This project involves the design of a fixed single hemispherical reflective surface to serve as a collector of solar energy. The hemispherical surface will reflect the sun's rays onto a boiler which will, in turn, provide hot water or steam as a source of energy.

Background information and suggested approach

The hemisphere is to be about 15 m in diameter, and to be lined with 0.75 mm thick aluminum sheet purchased in rolls 2 m wide and 30 m long. This material weighs 2.1 kg/m², or 126 kg/roll.

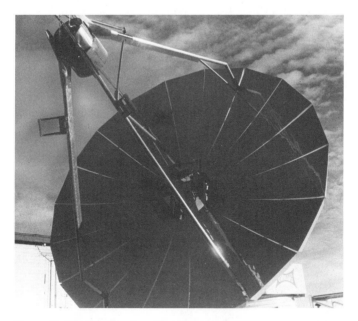

Figure 10–39   The Omnium-G solar collector. (*Courtesy* Solar Energy Research Institute.)

The hemispherical collector is stationary, and its construction is consequently less expensive than that of a movable reflective surface. Since a hemisphere does not have an axis but is symmetrical about a point, it does not have to be aimed toward the sun, as a parabolic collector would. However, a hemispherical reflecting surface does not focus parallel rays onto a point as a parabolic surface does, and therefore is not as efficient. A stationary collector is less expensive and is less complex. The boiler is not a point, but a volume, so point focus is not absolutely necessary.

Assume the boiler is about 1 m³ in volume, suspended by a boom or by cables. The boiler may be moved seasonally, but not hourly or daily.

The first main part of this project is to determine how the spherical surface will be fabricated. Decide what shape the pieces will be, how many are needed, how much material must be purchased, and how much of this material will be wasted. Drawings and calculations should be shown. Since a point focus is not possible with this shape of reflector, the surface can approximate a sphere. For example, small flat, conical, or cylindrical areas could form a part of the surface. Estimate the total seam length.

The second main part of this project is to determine graphically the shape and position of the boiler such that the maximum amount of sunlight will be absorbed by the boiler. Your hemisphere may be built in a horizontal position or tilted toward the south. Project 10.9 contains some equations that will allow you to predict approximately the angle of approach of the sun's rays if you know the time of day, day of the year, and your latitude.

Figure 10–40 shows an analysis of seven parallel rays (all in a vertical plane through the center of the sphere) with their pattern of reflection. This is a simplification, since the rays not in this plane will be reflected in three-dimensional space, and not just two dimensional as shown in this drawing.

Figure 10–41 shows a single entering ray, AB, hitting the reflective surface at point B. The reflected ray BC will make an equal angle with OB, the normal to the surface. Ray BC will lie in the extended plane ABO. Thus, auxiliary view 1 is drawn in the direction OX off the top view, showing plane OBA as an edge. Auxiliary view 2 shows the true shape of plane OBA. In this view, BC is constructed so that angle ABO = angle OBC. Point C is then transferred back from view 2 to view 1, and finally into the top and front views. Reference line H/3 is drawn parallel to BC in the top view, and auxiliary view 3 is drawn in order to determine the possible intersection of line BC and the spherical surface. View 3 shows line BC intersecting the surface at point Y.

In order to find the direction of the reflected ray, additional auxiliary views would have to be drawn showing the edge view and true shape of plane BYO. Clearly, it is a long, tedious process to plot the path of a single ray.

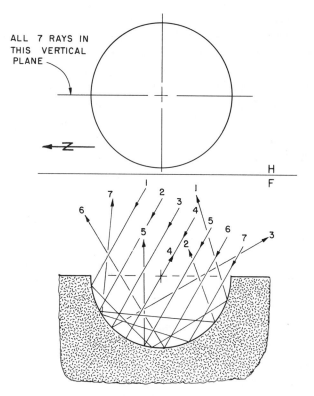

ALL 7 RAYS IN
THIS VERTICAL
PLANE

Figure 10–40 Reflection of parallel rays in a circular mirror.

Required output

Your written report should include drawings showing the position of the hemisphere and the boiler, how the boiler is supported, and how it can be relocated, if necessary, for seasonal adjustment. Try to estimate what percent of sunlight entering the mouth of the collector will hit the boiler.

10.12 Project 12: Quality control

Quality control is a very important part of the manufacturing process. Industries have devised various means for testing their products (see Figure 10–42). The purpose of this project is to design a testing program for examining wear and possible failure of an oven door. A robot will be used to perform the test.

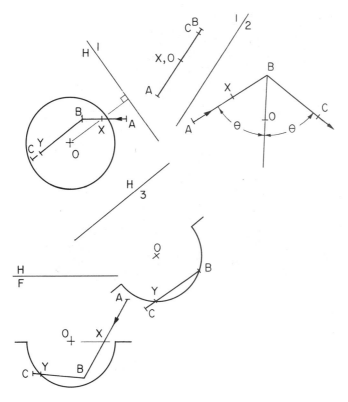

Figure 10–41   Using three auxiliary views to graphically determine the reflected path of a light ray entering a spherical mirror.

Background information and suggested approach

Figure 10–43 shows an oven with a rectangular door hinged along the bottom edge. This door is spring-loaded so it will fall to a horizontal position if it is opened 30° or more. Similarly, if the door is pulled to within 30° of vertical, the spring will close the door.

The door handle is fastened to the door with two machine screws. The metric threads are M6. This handle will probably be removed for the experimental test, and an appropriate testing device will be fastened to the oven door using these threads. (An M6 coarse thread has a 6-mm diameter and pitch of 1 mm/thread).

The test cycle for the door will consist of moving the door from its full open position to its full closed position and back 10,000 times. A robot will be used to perform the test. If the door is still functioning after 10,000 cycles, the hinge and spring will be examined for wear, and the test will continue until failure occurs.

Figure 10–42 Testing a vehicle for fuel system integrity. (*Courtesy* Chrysler Corporation.)

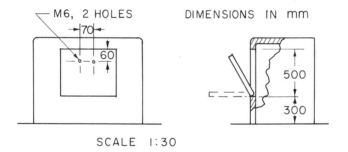

Figure 10–43 Dimensions for oven door and handle location.

The industrial robot R3D1, which is available for this project, is shown in Figure 10–44. Robot R3D1 has 3 angles of rotation (R3) and one arm extension distance (D1). All 3 angles, A, B, and C are driven by motors which operate at speeds of 3 rpm (or 18°/s) in a positive direction of rotation (shown in the drawing). If they move in the opposite direction, the rotation is negative. D can expand from 200 to 400 mm in length and moves at a velocity of 25 mm/s for both expansion and contraction.

Since opening and closing the door is a fairly simple task, this robot may make more types of motion than this task requires. Do not feel obligated to use all four controls (A, B, C, and D) if you do not need them; however, you must specify the initial position for all four.

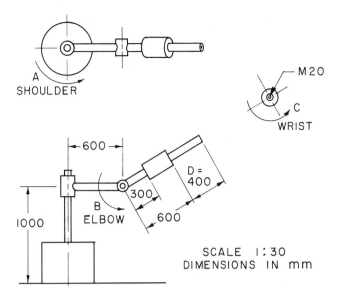

Figure 10–44   Dimensions of robot R3D1.

Required output

The project objectives are twofold. First, something equivalent to a hand and handle must be devised so the robot can move the door. Second, a work schedule must be carefully planned to operate the robot through a complete cycle. The end of the robot's arm has an internal M20 thread, so whatever is devised to attach to this arm can simply be screwed into this tapped hole. Your report should include scale drawings of the hand and handle used to open and close the door.

In order to plan the movement for the robot, you must first establish its location in the room with respect to the oven. A drawing showing this location could also help establish the initial positions. For example, what position is the robot in when A = 0, B = 0, C = 0, and D = 0?

A complete work cycle must include information for all four controls for the entire time period. An example is shown in Figure 10–45 for rotation A of the shoulder. In this illustration the operation of A (the shoulder) is shown for the first 5 seconds of a cycle. A is turned on after 1 second, then turned to reverse at t = 3, and turned off at t = 4. The resulting rotation is plotted as a function of time, showing A rotates 18 deg/s with the net result of a +18° rotation after 5 seconds.

After the cycle is complete and the robot is back in its original position, the time per cycle will be established, and the total time for 10,000 cycles can be computed. Cycles for all four controls should be included in your final report.

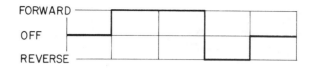

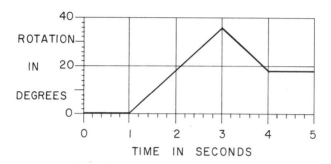

Figure 10–45 A typical work cycle for one angle of rotation.

Figure 10–46 An industrial robot at work on an automobile assembly line. (*Courtesy* Unimation, Inc.)

   An approximate scale model of this robot could be constructed from material similar to Tinker Toys.

   Your report could discuss the following points:

1. Could a less sophisticated robot be used as effectively for this test? What features do you feel a robot should include which are not on this robot?
2. How might a safety device be designed which would turn off the robot if the door falls off the oven (failure)?
3. What other types of tasks could this robot perform? What types of tasks could not be performed by this robot?

Project plan

   Locate the robot with respect to the stove. Prepare a complete work cycle for all four controlled motions. Design an appropriate device for attaching the robot's arm to the oven door. Discuss results and suggested additional items. (Figure 10–46 shows a typical industrial robot at work.)

10.13   Project 13: Packaging design

Project statement

   The purpose of this project is to design a means of packaging the computer terminal shown in Figure 10–47.

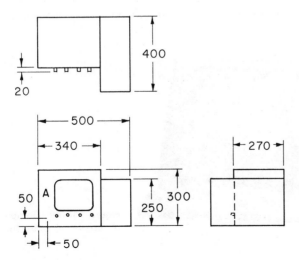

Figure 10–47   Three orthographic views of a computer terminal.

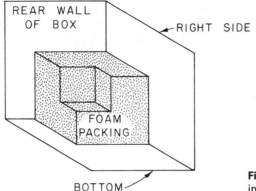

Figure 10–48 Packaging for protecting a corner.

Background information

Assume this terminal and its protective padding will be placed in a corrugated cardboard box of a rectangular prism shape. The object must be kept at least 30 mm away from the walls of the shipping box to prevent damage. The knobs and screen on surface A should not come in contact with any packaging. Packaging can touch the 50-mm strip along the left and bottom edges of this surface.

Required output

Sketch or draw your design for the soft foam plastic protective packaging material so that it may be fabricated. Specify how many pieces of each shape are

Figure 10–49 Computer terminal. (*Courtesy* Tektronix, Inc.)

needed per box. You may use either multiview or pictorial drawings, or a combination of these two. Show how the plastic protectors are to be positioned in the box. Number your pieces, and give instructions explaining the order in which the pieces and the objects are to be placed in the box. You may have between two and eight pieces of protective foam.

Compute the total volume of the packaging and the overall dimensions of the box. A sample drawing of a possible packaging shape is shown in Figure 10–48, and a typical computer terminal is shown in Figure 10–49.

REFERENCES

ALGER, J. R., and C. V. HAYES, *Creative Synthesis in Design*. Englewood Cliffs, N. J.: Prentice-Hall, 1964.

ASIMOW, M., *Introduction to Design*. Englewood Cliffs, N. J.: Prentice-Hall, 1962.

"Fixed Mirrors to Power City," *Machine Design,* vol. 48, no. 23, Oct. 7, 1976, p. 6.

GIBSON, J. E., *Introduction to Engineering Design*. New York: Holt, Rinehart and Winston, 1968.

HILL, P. H., *The Science of Engineering Design*. New York: Holt, Rinehart and Winston, 1970.

KRICK, E. V., *An Introduction to Engineering and Engineering Design*, 2nd ed. New York: Wiley, 1969.

"Power Tower Progress Report" *Machine Design*, vol. 50, no. 8, April 6, 1978, p. 18.

"Solar Energy, the Ultimate Powerhouse," *National Geographic,* John L. Wilhelm, vol. 149, no. 3, March, 1976, pp. 381–397

.SPOTTS, M. F., *Design Engineering Projects*. Englewood Cliffs, N. J.: Prentice-Hall, 1968.

Screw threads
and fasteners
A

A-1 Metric Screw Thread Profile (ANSI B1.13M–1979)

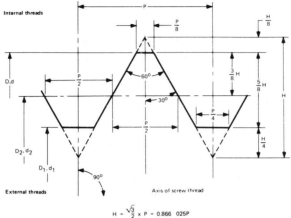

$$H = \frac{\sqrt{3}}{2} \times P = 0.866\ 025P$$

$0.125H = 0.108\ 253P$
$0.250H = 0.216\ 506P$
$0.375H = 0.324\ 760P$
$0.625H = 0.541\ 266P$

BASIC M THREAD PROFILE (ISO 68 BASIC PROFILE)

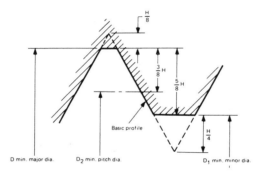

INTERNAL THREAD, DESIGN M PROFILE (MAXIMUM MATERIAL CONDITION)

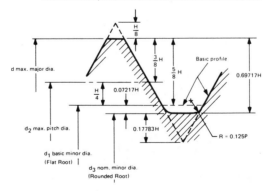

EXTERNAL THREAD, DESIGN M PROFILE WITH NO ALLOWANCE (FUNDAMENTAL
DEVIATION) (FLANKS AT MAXIMUM MATERIAL CONDITION)

A–2 Metric Thread Designation (ANSI B1.13M–1979)

5.1 General

5.1.1 The complete designation of a screw thread gives the thread symbol, the nominal size and the thread tolerance class.

5.1.2 The tolerance class designation gives the class designation for the pitch diameter tolerance followed by a class designation for the crest diameter (major diameter for external thread and minor diameter for internal thread) tolerances.

5.1.3 The class designation consists of a number indicating the tolerance grade followed by a letter indicating the tolerance position.

Examples:

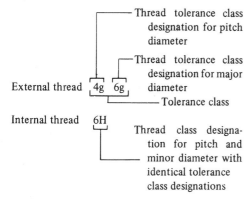

5.2 Designation of Standard Screw Threads

Metric screw threads are identified by letter (M) for the thread form profile, followed by the nominal diameter size and the pitch expressed in millimeters, separated by the sign (x) and followed by the tolerance class separated by a dash (–) from the pitch.

The simplified international practice for designating coarse pitch M profile screw threads is to leave off the pitch. Thus a M14 x 12 thread is designated just M14. To prevent misunderstanding, it is mandatory to use the value for pitch in all designations.

5.2.1 Unless otherwise specified in the designation, the screw thread helix is right hand.

Example:

External Thread M Profile, Right Hand:

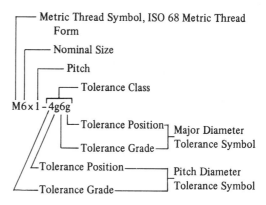

Internal Thread M Profile, Right Hand:

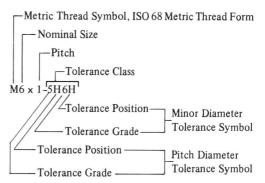

5.2.2 Designation of Left Hand Thread. When left hand thread is specified, the tolerance class designation is followed by a space and LH.

Example:

M6x1–5H6H–LH

A–3 Metric Screw Thread Series (ANSI B1.13M–1979)

Nominal Diameters			Pitches										
Col. 1 1st Choice	Col. 2 2nd Choice	Col. 3 3rd Choice	Coarse	Fine									
				3	2	1.5	1.25	1	0.75	0.5	0.35	0.25	0.2
1.6			0.35	...	...	...	...	...	...	...	...	...	0.2
	1.8		0.35	...	...	...	...	...	...	...	...	...	0.2
2			0.4	...	...	...	...	...	...	...	...	0.25	...
	2.2		0.45	...	...	...	...	...	...	...	...	0.25	...
2.5			0.45	...	...	...	...	...	...	...	0.35	...	...
3			0.5	...	...	...	...	...	...	...	0.35	...	...
	3.5		0.6	...	...	...	...	...	...	...	0.35	...	...
4			0.7	...	...	...	...	...	...	0.5	...	...	...
	4.5		0.75	...	...	...	...	...	...	0.5	...	...	...
5			0.8	...	...	...	...	...	...	0.5	...	...	...
		5.5	...	...	...	...	...	...	...	0.5	...	...	...
6			1	...	...	...	...	...	0.75	...	...	...	...
		7	1	...	...	...	...	...	0.75	...	...	...	...
8			1.25	...	...	...	...	1	0.75	...	...	...	...
		9	1.25	...	...	...	...	1	0.75	...	...	...	...
10			1.5	...	...	...	1.25	1	0.75	...	...	...	...
		11	1.5	...	...	...	...	1	0.75	...	...	...	...
12			1.75	...	...	1.5	1.25	1	...	...	...	...	...
	14		2	...	...	1.5	1.25	1	...	...	...	...	...
		15	...	...	...	1.5	...	1	...	...	...	...	...
16			2	...	...	1.5	...	1	...	...	...	...	...
		17	...	...	...	1.5	...	1	...	...	...	...	...
	18		2.5	...	2	1.5	...	1	...	...	...	...	...
20			2.5	...	2	1.5	...	1	...	...	...	...	...
	22		2.5	...	2	1.5	...	1	...	...	...	...	...
24			3	...	2	1.5	...	1	...	...	...	...	...
		25	...	...	2	1.5	...	1	...	...	...	...	...
		26	...	...	...	1.5	...	...	...	...	...	...	...
	27		3	...	2	1.5	...	1	...	...	...	...	...
		28	...	...	2	1.5	...	1	...	...	...	...	...
30			3.5	(3)	2	1.5	...	1	...	...	...	...	...
	32		...	...	2	1.5	...	...	...	...	...	...	...
	33		3.5	(3)	2	1.5	...	...	...	...	...	...	...
		35	...	...	...	1.5	...	...	...	...	...	...	...
36			4	3	2	1.5	...	...	...	...	...	...	...
		38	...	...	...	1.5	...	...	...	...	...	...	...
	39		4	3	2	1.5	...	...	...	...	...	...	...

A–3 (continued)

Nominal Diameters			Pitches					
Col. 1 1st Choice	Col. 2 2nd Choice	Col. 3 3rd Choice	Coarse	Fine				
				6	4	3	2	1.5
		40	...	...	...	3	2	1.5
42			4.5	...	4	3	2	1.5
	45		4.5	...	4	3	2	1.5
48			5	...	4	3	2	1.5
		50	...	...	...	3	2	1.5
	52		5	...	4	3	2	1.5
		55	...	...	4	3	2	1.5
56			5.5	...	4	3	2	1.5
		58	...	...	4	3	2	1.5
	60		5.5	...	4	3	2	1.5
		62	...	...	4	3	2	1.5
64			6	...	4	3	2	1.5
		65	...	...	4	3	2	1.5
	68		6	...	4	3	2	1.5
		70	...	6	4	3	2	1.5
72			6	6	4	3	2	1.5
		75	...	...	4	3	2	1.5
	76		...	6	4	3	2	1.5
		78	...	...	...	...	2	...
80			6	6	4	3	2	1.5
		82	...	...	...	...	2	...
	85		...	6	4	3	2	...
90			6	6	4	3	2	...
	95		...	6	4	3	2	...
100			6	6	4	3	2	...
	105		...	6	4	3	2	...
110			...	6	4	3	2	...
	115		...	6	4	3	2	...
	120		...	6	4	3	2	...
125			...	6	4	3	2	...
	130		...	6	4	3	2	...
		135	...	6	4	3	2	...
140			...	6	4	3	2	...
		145	...	6	4	3	2	...
	150		...	6	4	3	2	...
		155	...	6	4	3	...	...
160			...	6	4	3	...	...
		165	...	6	4	3	...	...
	170		...	6	4	3	...	...
		175	...	6	4	3	...	...
180			...	6	4	3	...	...
		185	...	6	4	3	...	...
	190		...	6	4	3	...	...
		195	...	6	4	3	...	...
200			...	6	4	3	...	...

A–3 (continued)

Nominal Diameters			Pitches					
Col. 1 1st Choice	Col. 2 2nd Choice	Col. 3 3rd Choice	Coarse	Fine				
				6	4	3	2	1.5
		205	. . .	6	4	3	. . .	. . .
	210		. . .	6	4	3	. . .	. . .
		215	. . .	6	4	3	. . .	. . .
220			. . .	6	4	3	. . .	. . .
		225	. . .	6	4	3	. . .	. . .
		230	. . .	6	4	3	. . .	. . .
		235	. . .	6	4	3	. . .	. . .
	240		. . .	6	4	3	. . .	. . .
		245	. . .	6	4	3	. . .	. . .
250			. . .	6	4	3	. . .	. . .
		255	. . .	6	4	. . .	. . .	. . .
	260		. . .	6	4	. . .	. . .	. . .
		265	. . .	6	4	. . .	. . .	. . .
		270	. . .	6	4	. . .	. . .	. . .
		275	. . .	6	4	. . .	. . .	. . .
280			. . .	6	4	. . .	. . .	. . .
		285	. . .	6	4	. . .	. . .	. . .
		290	. . .	6	4	. . .	. . .	. . .
		295	. . .	6	4	. . .	. . .	. . .
	300		. . .	6	4	. . .	. . .	. . .

A–4 Unified Thread Designation (ANSI Y14.6–1978)

In general practice the general designation and the pitch diameter limits are in note form and referenced to the drawing of the thread with a leader line. The following example illustrates and explains the elements of a designation of the screw thread.

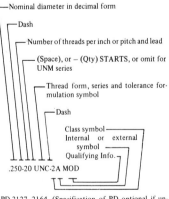

—Nominal diameter in decimal form

—Dash

—Number of threads per inch or pitch and lead

— (Space), or – (Qty) STARTS, or omit for UNM series

—Thread form, series and tolerance formulation symbol

—Dash

Class symbol
Internal or external symbol
Qualifying Info.

.250-20 UNC-2A MOD

PD.2127-.2164 (Specification of PD optional if uncoated)

A–5 Unified Standard Screw Thread Series (ANSI Y14.6–1978)

Primary	Secondary	BASIC MAJOR DIAMETER	Coarse UNC	Fine UNF	Extra fine UNEF	4UN	6UN	8UN	12UN	16UN	20UN	28UN	32UN	SIZES
0		0.060	–	80	–	–	–	–	–	–	–	–	–	0
	1	0.073	64	72	–	–	–	–	–	–	–	–	–	1
2		0.086	56	64	–	–	–	–	–	–	–	–	–	2
	3	0.099	48	56	–	–	–	–	–	–	–	–	–	3
4		0.112	40	48	–	–	–	–	–	–	–	–	–	4
5		0.125	40	44	–	–	–	–	–	–	–	–	–	5
6		0.138	32	40	–	–	–	–	–	–	–	–	UNC	6
8		0.164	32	36	–	–	–	–	–	–	–	–	UNC	8
10		0.190	24	32	–	–	–	–	–	–	–	–	UNF	10
	12	0.216	24	28	32	–	–	–	–	–	–	UNF	UNEF	12
1/4		0.250	20	28	32	–	–	–	–	–	UNC	UNF	UNEF	1/4
5/16		0.3125	18	24	32	–	–	–	–	–	20	28	UNEF	5/16
3/8		0.375	16	24	32	–	–	–	–	UNC	20	28	UNEF	3/8
7/16		0.4375	14	20	28	–	–	–	–	16	UNF	UNEF	32	7/16
1/2		0.500	13	20	28	–	–	–	–	16	UNF	UNEF	32	1/2
9/16		0.5265	12	18	24	–	–	–	UNC	16	20	28	32	9/16
5/8		0.625	11	18	24	–	–	–	12	16	20	28	32	5/8
	11/16	0.6875	–	–	24	–	–	–	12	16	20	28	32	11/16
3/4		0.750	10	16	20	–	–	–	12	UNF	UNEF	28	32	3/4
	13/16	0.8125	–	–	20	–	–	–	12	16	UNEF	28	32	13/16
7/8		0.875	9	14	20	–	–	–	12	16	UNEF	28	32	7/8
	15/16	0.9375	–	–	20	–	–	–	12	16	UNEF	28	32	15/16
1		1.000	8	12	20	–	–	UNC	UNF	16	UNEF	28	32	1
	1 1/16	1.0625	–	–	18	–	–	8	12	16	20	28	–	1 1/16
1 1/8		1.125	7	12	18	–	–	8	UNF	16	20	28	–	1 1/8
	1 3/16	1.1875	–	–	18	–	–	8	12	16	20	28	–	1 3/16
1 1/4		1.250	7	12	18	–	–	8	UNF	16	20	28	–	1 1/4
	1 5/16	1.3125	–	–	18	–	–	8	12	16	20	28	–	1 5/16
1 3/8		1.375	6	12	18	–	UNC	8	UNF	16	20	28	–	1 3/8
	1 7/16	1.4375	–	–	18	–	6	8	12	16	20	28	–	1 7/16
1 1/2		1.500	6	12	18	–	UNC	8	UNF	16	20	28	–	1 1/2
	1 9/16	1.5625	–	–	18	–	6	8	12	16	20	–	–	1 9/16
1 5/8		1.625	–	–	18	–	6	8	12	16	20	–	–	1 5/8
	1 11/16	1.6875	–	–	18	–	6	8	12	16	20	–	–	1 11/16
1 3/4		1.750	5	–	–	–	6	8	12	16	20	–	–	1 3/4
	1 13/16	1.8125	–	–	–	–	6	8	12	16	20	–	–	1 13/16
1 7/8		1.875	–	–	–	–	6	8	12	16	20	–	–	1 7/8
	1 15/16	1.9375	–	–	–	–	6	8	12	16	20	–	–	1 15/16
2		2.000	4 1/2	–	–	–	6	8	12	16	20	–	–	2
	2 1/8	2.125	–	–	–	–	6	8	12	16	20	–	–	2 1/8
2 1/4		2.250	4 1/2	–	–	–	6	8	12	16	20	–	–	2 1/4
	2 3/8	2.375	–	–	–	–	6	8	12	16	20	–	–	2 3/8
2 1/2		2.500	4	–	–	UNC	6	8	12	16	20	–	–	2 1/2
	2 5/8	2.625	–	–	–	4	6	8	12	16	20	–	–	2 5/8
2 3/4		2.750	4	–	–	UNC	6	8	12	16	20	–	–	2 3/4
	2 7/8	2.875	–	–	–	4	6	8	12	16	20	–	–	2 7/8
3		3.000	4	–	–	UNC	6	8	12	16	20	–	–	3
	3 1/8	3.125	–	–	–	4	6	8	12	16	–	–	–	3 1/8
3 1/4		3.250	4	–	–	UNC	6	8	12	16	–	–	–	3 1/4
	3 3/8	3.375	–	–	–	4	6	8	12	16	–	–	–	3 3/8
3 1/2		3.500	4	–	–	UNC	6	8	12	16	–	–	–	3 1/2
	3 5/8	3.625	–	–	–	4	6	8	12	16	–	–	–	3 5/8
3 3/4		3.750	4	–	–	UNC	6	8	12	16	–	–	–	3 3/4
	3 7/8	3.875	–	–	–	4	6	8	12	16	–	–	–	3 7/8
4		4.000	4	–	–	UNC	6	8	12	16	–	–	–	4
	4 1/8	4.125	–	–	–	4	6	8	12	16	–	–	–	4 1/8
4 1/4		4.250	–	–	–	4	6	8	12	16	–	–	–	4 1/4
	4 3/8	4.375	–	–	–	4	6	8	12	16	–	–	–	4 3/8
4 1/2		4.500	–	–	–	4	6	8	12	16	–	–	–	4 1/2
	4 5/8	4.625	–	–	–	4	6	8	12	16	–	–	–	4 5/8
4 3/4		4.750	–	–	–	4	6	8	12	16	–	–	–	4 3/4
	4 7/8	4.875	–	–	–	4	6	8	12	16	–	–	–	4 7/8
5		5.000	–	–	–	4	6	8	12	16	–	–	–	5
	5 1/8	5.125	–	–	–	4	6	8	12	16	–	–	–	5 1/8
5 1/4		5.250	–	–	–	4	6	8	12	16	–	–	–	5 1/4
	5 3/8	5.375	–	–	–	4	6	8	12	16	–	–	–	5 3/8
5 1/2		5.500	–	–	–	4	6	8	12	16	–	–	–	5 1/2
	5 5/8	5.625	–	–	–	4	6	8	12	16	–	–	–	5 5/8
5 3/4		5.750	–	–	–	4	6	8	12	16	–	–	–	5 3/4
	5 7/8	5.875	–	–	–	4	6	8	12	16	–	–	–	5 7/8
6		6.000	–	–	–	4	6	8	12	16	–	–	–	6

A–6 Dimensions of Carriage Bolt with Hex Nut* (Grade 5.6, Coarse Thread)

Length Available	Nominal Diameter of Bolt					
	M5	M6	M8	M10	M12	M16
16	X	X				
20	X	X	X			
25	X	X	X	X		
30	X	X	X	X	X	
35	X	X	X	X	X	X
40	X	X	X	X	X	X
45	X	X	X	X	X	X
50	X	X	X	X	X	X
55		X	X	X	X	X
60		X	X	X	X	X
65		X	X	X	X	X
70		X	X	X	X	X
75		X	X	X	X	X
80		X	X	X	X	X
90		X	X	X	X	X
100		X	X	X	X	X
110			X	X	X	X
120			X	X	X	X
130			X	X	X	X
140			X	X	X	X
150				X	X	X
160				X	X	X
180					X	
200					X	

*All dimensions in millimeters.

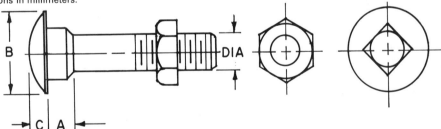

Nominal Dia.		M5	M6	M8	M10	M12	M16
Pitch	mm	0.8	1	1.25	1.5	1.75	2
Squares	mm	5×5	6×6	8×8	10×10	12×12	16×16
A	mm	3.5	4	5	6	8	12
B	mm	13	16	20	24	30	38
C	mm	3	3.5	4.5	5	6.5	8.5
Thread Length							
under 125 mm		16	18	22	26	30	38
over 125 mm		22	24	28	32	36	44
nut across flats	mm	8	10	13	17	19	24

A–7 Dimensions of Socket Head Cap Screws*

Thread					Screw Dimensions									
		Screw Length			D		K		S			A		
Nom. Dia.	Pitch	under 125	130 200	over 200	Max.	Min.	Max.	Min.	Nom.	Max.	Min.	Max.	Min.
		Thread Length											
M 1.4	0.3				2.6		1.4		1.27			0.8	0.6
M 1.6	0.35				3		1.7		1.5			0.9	0.7
M 1.7	0.35				3		1.7		1.5				
M 2	0.4				3.5		2		1.5			1.3	1
M 2.3	0.4				4		2.3		2				
M 2.5	0.45				4.5		2.5		2			1.5	1.1
M 2.6	0.45				4.5		2.6		2				
M 3	0.5	12			5.5	5.32	3	2.86	2.5	2.62	2.52	1.7	1.3
M 4	0.7	14			7	6.78	4	3.82	3	3.12	3.02	2.4	2
M 5	0.8	16			8.5	8.28	5	4.82	4	4.15	4.03	3.1	2.7
M 6	1	18	24		10	9.78	6	5.82	5	5.15	5.03	3.78	3.3
M 8	1.25	22	28		13	12.73	8	7.78	6	6.15	6.03	4.78	4.3
M 10	1.5	26	32	45	16	15.73	10	9.78	8	8.19	8.04	6.25	5.5
M 12	1.75	30	36	49	18	17.73	12	11.73	10	10.19	10.04	7.5	6.6
M 14	2	34	40	53	21	20.67	14	13.73	12	12.23	12.05	8.7	7.8
M 16	2	38	44	57	24	23.67	16	15.73	14	14.23	14.05	9.7	8.8
M 18	2.5	42	48	61	27	26.67	18	17.73	14	14.23	14.05	10.7	9.8
M 20	2.5	46	52	65	30	29.67	20	19.67	17	17.23	17.05	11.8	10.7
M 22	2.5	50	56	69	33	32.61	22	21.67	17	17.23	17.05	12.4	11.3
M 24	3	54	60	73	36	35.61	24	23.67	19	19.28	19.07	14	12.9
M 27	3	60	66	79	40	39.61	27	26.67	19	19.29	19.07	16.2	15.1
M 30	3.5	66	72	85	45	44.61	30	29.67	22	22.28	22.07	18.2	17.1
M 33	3.5	72	78	91	50	49.61	33	32.61	24	24.28	24.07	20.1	18.8
M 36	4	78	84	97	54	53.54	36	35.61	27	27.28	27.07	22.1	20.8
M 42	4.5	90	96	109	63	62.54	42	41.61	32	32.33	32.08	26.3	25.0
M 48	5	102	108	121	72	71.54	48	47.61	36	36.33	36.08	30.4	29.1

*All dimensions in millimeters.

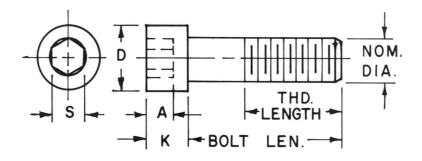

A–8 Dimensions of Hex Head Cap Screws*

Thread					Bolt			
Nominal Diameter	Pitch	Bolt Length			S	K	C	E
		under 125	125-200	over 200				
		Thread Length						
M 1	0.25							
M 1.2	0.25							
M 1.4	0.3							
M 1.6	0.35	9			3.2	1.1		3.48
M 1.7	0.35	9			3.5	1.2		3.82
M 1.8	0.35							
M 2	0.4	10			4	1.4		4.38
M 2.3	0.4	11			4.5	1.6		4.95
M 2.5	0.45	11			5	1.7		5.51
M 2.6	0.45	11			5	1.8		5.51
M 3	0.5	12			5.5	2		6.08
M 3.5	0.6	13			6	2.4		6.64
M 4	0.7	14			7	2.8	0.1	7.74
M 5	0.8	16	22		8	3.5	0.2	8.87
M 6	1	18	24		10	4	0.3	11.05
M 7	1	20	26		11	5	0.3	12.12
M 8	1.25	22	28		13	5.5	0.4	14.38
M 10	1.5	26	32	45	17	7	0.4	18.90
M 12	1.75	30	36	49	19	8	0.4	21.10
M 14	2	34	40	53	22	9	0.4	24.49
M 16	2	38	44	57	24	10	0.4	26.75
M 18	2.5	42	48	61	27	12	0.4	30.14
M 20	2.5	46	52	65	30	13	0.4	33.53
M 22	2.5	50	56	69	32	14	0.4	35.72
M 24	3	54	60	73	36	15	0.5	39.98
M 27	3	60	66	79	41	17	0.5	45.63
M 30	3.5	66	72	85	46	19	0.5	51.28
M 33	3.5	72	78	91	50	21	0.5	55.80
M 36	4	78	84	97	55	23	0.5	61.31

*All dimensions in millimeters

A–8 (continued)

Thread						Bolt			
Nominal Diameter	Pitch	Bolt Length			S	K	C	E
		under 125	125-200	over 200				
		Thread Length						
M 39	4	84	90	103	60	25	0.6	66.96
M 42	4.5	90	96	109	65	26	0.6	72.61
M 45	4.5	96	102	115	70	28	0.6	78.26
M 48	5	102	108	121	75	30	0.6	83.91
M 52	5		116	129	80	33		89.56
M 56	5.5		124	137	85	35		95.07
M 60	5.5		132	145	90	38		100.72
M 64	6		140	153	95	40		106.37
M 68	6		148	161	100	43		112.02
M 72	6		156	169	105	45		117.67
M 76	6		164	177	110	48		123.32
M 80	6		172	185	115	50		128.97
M 90	6		192	205	130	57		145.77
M 100	6			225	145	63		162.72
M 110	6			245	155	69		174.02
M 120	6			265	170	76		190.97
M 125	6			275	180	79		202.27
M 130	6			285	185	82		207.75
M 140	6			305	200	88		224.70
M 150	6			325	210	95		236.00

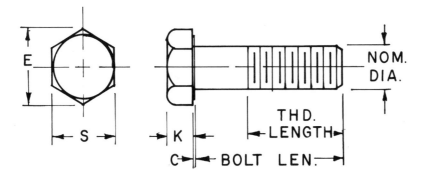

A–9 Dimensions of Hex Nuts*

Hexagonal Nuts					
Thread				Nut	
	Pitch				
Nom. Dia.	Coarse		Fine	S	M
M 1	0.25			2.5	0.8
M 1.2	0.25			3	1
M 1.4	0.3			3	1.2
M 1.6	0.35			3.2	1.3
M 1.7	0.35			3.5	1.4
M 2	0.4			4	1.6
M 2.3	0.4			4.5	1.8
M 2.5	0.45			5	2
M 2.6	0.45			5	2
M 3	0.5			5.5	2.4
M 3.5	0.6			6	2.8
M 4	0.7			7	3.2
M 5	0.8			8	4
M 6	1	0.75	0.5	10	5
M 7	1			11	5.5
M 8	1.25	1		13	6.5
M 10	1.5	1.25	1	17	8
M 12	1.75	1.5	1.25	19	10
M 14	2	1.5		22	11
M 16	2	1.5		24	13
M 18	2.5	2	1.5	27	15
M 20	2.5	2	1.5	30	16
M 22	2.5	2	1.5	32	18
M 24	3	2	1.5	36	19
M 26			1.5	41	22
M 27	3	2	1.5	41	22
M 28			1.5	41	22
M 30	3.5	2	1.5	46	24
M 32			1.5	50	26
M 33	3.5	2	1.5	50	26
M 35			1.5	55	29

Hexagonal Nuts					
Thread				Nut	
	Pitch				
Nom. Dia.	Coarse		Fine	S	M
M 36	4	3	1.5	55	29
M 38			1.5	60	31
M 39	4	3	1.5	60	31
M 40			1.5	60	31
M 42	4.5	3	1.5	65	34
M 45	4.5	3	1.5	70	36
M 48	5	3	1.5	75	38
M 50			1.5	75	38
M 52	5	3	1.5	80	42
M 56	5.5	4	1.5	85	45
M 58			2	90	48
M 60	5.5	4	2	90	48
M 64	6	4	2	95	51
M 68	6	4	2	100	54
M 72	6	4	2	105	58
M 76	6	4	2	110	61
M 80	6	4	2	115	64
M 85	6	4	2	120	68
M 90	6	4	2	130	72
M 95	6	4	2	135	76
M 100	6	4	2	145	80
M 105	6	4	2	150	84
M 110	6	4	2	155	88
M 115	6	4	2	165	92
M 120	6	4	2	170	96
M 125	6	4	2	180	100
M 130	6		3	185	104
M 135	6		3	190	108
M 140	6		3	200	112
M 145	6		3	210	116
M 150	6		3	210	120

*All dimensions are in millimeters

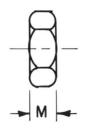
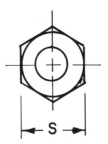

A–10 Dimensions of Flat Washers*

Nom. Size I.D.	O.D.	T	Screw Dia.
1.7	4	0.3	1.6
2.2	5	0.3	2
2.7	6.5	0.5	2.5
3.2	7	0.5	3
4.3	9	0.8	4
5.3	10	1	5
6.4	12.5	1.6	6
7.4	14	1.6	7
8.4	17	1.6	8
10.5	21	2	10
13	24	2.5	12
15	28	2.5	14
17	30	3	16
19	34	3	18
21	37	3	20
23	39	3	22
25	44	4	24
28	50	4	27
31	56	4	30
34	60	5	33
37	66	5	36
40	72	6	39
43	78	7	42
46	85	7	45
50	92	8	48
54	98	8	52
58	105	9	56
62	110	9	60
66	115	9	64
70	120	10	68

*All dimensions in millimeters.

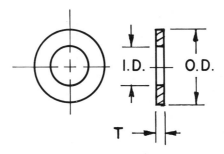

A–11 Dimensions of Twist Drills.*

Tolerance h8

Size Dia.	Overall Length	Size Dia.	Overall Length	Size Dia.	Overall Length	Size Dia.	Overall Length	Size Dia.	Overall Length
0.25	19	2.5	56	5.5	91	9.2	122	13.75	156
0.3	19	2.55	56	5.6	91	9.25	122	13.8	156
0.35	19	2.6	56	5.7	91	9.3	122	14	156
0.4	20	2.65	56	5.75	91	9.4	122	14.25	165
0.45	20	2.7	60	5.8	91	9.5	122	14.5	165
0.5	22	2.75	60	5.9	91	9.6	130	14.75	165
0.55	24	2.8	60	6	91	9.7	130	15	165
0.6	24	2.85	60	6.1	99	9.75	130	15.25	173
0.65	26	2.9	60	6.2	99	9.8	130	15.5	173
0.7	28	2.95	60	6.25	99	9.9	130	15.75	173
0.75	28	3	60	6.3	99	10	130	16	173
0.8	30	3.1	64	6.4	99	10.1	130	16.25	190
0.85	30	3.15	64	6.5	99	10.2	130	16.5	190
0.9	32	3.2	64	6.6	99	10.3	130	16.75	190
0.95	32	3.25	64	6.7	99	10.4	130	17	190
1	34	3.3	64	6.75	107	10.5	130	17.25	190
1.05	34	3.35	69	6.8	107	10.7	139	17.5	203
1.15	36	3.4	69	6.9	107	10.8	139	17.75	203
1.2	38	3.45	69	7	107	10.9	139	18	203
1.25	38	3.5	69	7.1	107	11	139	18.25	203
1.3	38	3.55	69	7.2	107	11.1	139	18.5	203
1.35	40	3.6	69	7.25	107	11.2	139	18.75	203
1.4	40	3.65	69	7.3	107	11.3	139	19	203
1.45	40	3.7	69	7.4	107	11.4	139	19.25	203
1.5	40	3.8	74	7.5	107	11.5	139	19.5	203
1.55	43	3.9	74	7.6	114	11.6	139	19.75	203
1.6	43	4	74	7.7	114	11.7	139	20	203
1.65	43	4.1	74	7.75	114	11.8	139		
1.7	43	4.2	74	7.8	114	11.9	147		
1.75	45	4.25	74	7.9	114	12	147		
1.8	45	4.3	79	8	114	12.1	147		
1.85	45	4.4	79	8.1	114	12.2	147		
1.9	45	4.5	79	8.2	114	12.3	147		
1.95	48	4.6	79	8.25	114	12.4	147		
2	48	4.7	79	8.3	114	12.5	147		
2.05	48	4.75	79	8.4	114	12.6	147		
2.1	48	4.8	85	8.5	114	12.7	147		
2.15	52	4.9	85	8.6	122	12.8	147		
2.2	52	5	85	8.7	122	12.9	147		
2.25	52	5.1	85	8.75	122	13	147		
2.3	52	5.2	85	8.8	122	13.1	147		
2.35	52	5.25	85	8.9	122	13.2	147		
2.4	56	5.3	85	9	122	13.25	156		
2.45	56	5.4	91	9.1	122	13.5	156		

Tolerances†

Drill Dia.	Minus Tolerance
3	0.014
3–6	0.018
6–10	0.022
10–18	0.027
18–30	0.033
30–50	0.039
50–80	0.046

†Plus tolerance is zero

*All dimensions in millimeters. Twist drills are straight shank, jobber length, right hand, general purpose, regular helix, point angle 118°.

A–12 Dimensions of Rivets*

Nominal Size	Hole Size	Lengths Available	Round Head dia.	height	Flat Head dia.	depth
2	2.2	6,8,10,12,15	3.5	1.2	3.5	1.
3	3.2	6,8,10,12,15,18,20,22,25,30,32	5.2	1.8	5.2	1.5
4	4.3	6,8,10,12,15,18,20,22,25,30,32,35,40,45,50	7	2.4	7	2
5	5.3	8,10,12,15,18,20,22,25,30,32,35,40,45,50,55,60	8.8	3	8.8	2.5
6	6.4	10,12,15,18,20,22,25,30,32,35,40,45,50,55,60	10.5	3.6	10.5	3
8		10,12,15,18,20,22,25,30,32,35,40,45,50,55,60	14	4.8	14	4

*All dimensions in millimeters.

A–12 (continued)

Rivet Length (L) for Maximum Thickness of Joint (T)*

Rivet Dia.	Round Head						Flat Head					
	2	3	4	5	6	8	2	3	4	5	6	8
L	T max.						T max.					
6	2.5	2.5	1.5			.	3.5	3.5	3.5			
8	4	4	3	2.5			5	5	5	4.5		
10	6	6	5	4	3		7	7	7	6	5	
12	8	8	7	6	5	4	9	9	9	8	7	7
15	11	11	10	9	8	7	12	12	12	11	10	10
18		13	12	11	10	9		14	14	13	12	12
20		15	14	13	12	11		16	16	15	14	14
22		17	16	15	14	13		18	18	17	16	16
25		19	18	17	16	15		20	20	19	18	18
30		22	21	20	20	19		23	23	22	22	22
32		24	23	22	22	21		25	25	24	24	24
35		26	25	24	24	23		27	27	26	26	26
40		31	30	29	29	28		32	32	31	31	31
45			34	33	33	32			36	35	35	35
50			38	37	37	36			40	39	39	39
55				42	42	41				44	44	44
60				46	46	45				48	48	48

*All dimensions in millimeters.

A–13 Dimensions of regular hex keys*

Nominal Size (Across Flats)	Arm Length
0.7	30 × 7
0.9	30 × 12
1.3	40 × 15
1.5	50 × 15
2	55 × 16
2.5	60 × 20
3	65 × 20
4	72 × 25
5	80 × 28
6	90 × 32
7	95 × 34
8	100 × 36
9	105 × 40
10	112 × 40
11	130 × 42
12	125 × 45
14	140 × 55
17	160 × 60
19	175 × 70
22	200 × 80
24	224 × 90
27	250 × 100
30	280 × 112
32	315 × 125
36	355 × 140

*All dimensions in millimeters.

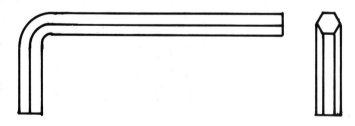

Constants and conversion factors

B

Table B-1 Physical Constants*

Quantity	Symbol	Value	×	Unit
Speed of light in vacuum	c	2.997 925	$\times 10^8$	m s^{-1}
Gravitational constant	G	6.673 2	$\times 10^{-11}$	N m^2 kg^{-2}
Avogadro constant	N_A	6.022 169	$\times 10^{26}$	kmol^{-1}
Boltzmann constant	k	1.380 622	$\times 10^{-23}$	J K^{-1}
Gas constant	R	8.314 34	$\times 10^3$	J kmol^{-1} K^{-1}
Faraday constant	F	9.648 670	$\times 10^7$	C kmol^{-1}
Planck constant	h	6.626 196	$\times 10^{-34}$	J s
Electron charge	e	1.602 191 7	$\times 10^{-19}$	C
Electron rest mass	m_e	9.109 558	$\times 10^{-31}$	kg
Proton rest mass	m_p	1.672 614	$\times 10^{-27}$	kg
Neutron rest mass	m_n	1.674 920	$\times 10^{-27}$	kg
Stefan-Boltzmann constant	σ	5.669 61	$\times 10^{-8}$	W m^{-2} K^{-4}
Rydberg constant	R_∞	1.097 373 12	$\times 10^7$	m^{-1}
Classical electron radius	r_e	2.817 939	$\times 10^{-15}$	m
Standard acceleration of free fall	g	9.806 65		m/s^2
Standard atmosphere	atm	101 325		Pa

* NASA SP-7012.

Table B-2 Conversion Factors†. (An asterisk follows each number which expresses an exact definition).

ALPHABETICAL LISTING

To convert from	to	multiply by
abampere	ampere	1.00* × 10
abcoulomb	coulomb	1.00* × 10
abfarad	farad	1.00* × 10^9
abhenry	henry	1.00* × 10^{-9}
abmho	siemens	1.00* × 10^9
abohm	ohm	1.00* × 10^{-9}
abvolt	volt	1.00* × 10^{-8}
acre	meter^2	4.046 856 422 4* × 10^3
angstrom	meter	1.00* × 10^{-10}
are	meter^2	1.00* × 10^2
astronomical unit (IAU)	meter	1.496 00 × 10^{11}
astronomical unit (radio)	meter	1.495 978 9 × 10^{11}
atmosphere	newton/meter^2	1.013 25* × 10^5

† NASA SP-7012, pp. 11–20.

Table B-2 *(continued)*

To convert from	to	multiply by
bar	newton/meter²	1.00* × 10⁵
barn	meter²	1.00* × 10⁻²⁸
barrel (petroleum, 42 gallons)	meter³	1.589 873 × 10⁻¹
barye	newton/meter²	1.00* × 10⁻¹
board foot (1′×1′×1″)	meter³	2.359 737 216* × 10⁻³
British thermal unit:		
(IST before 1956)	joule	1.055 04 × 10³
(IST after 1956)	joule	1.055 056 × 10³
British thermal unit (mean)	joule	1.055 87 × 10³
British thermal unit (thermochemical)	joule	1.054 350 × 10³
British thermal unit (39° F)	joule	1.059 67 × 10³
British thermal unit (60° F)	joule	1.054 68 × 10³
bushel (U.S.)	meter³	3.523 907 016 688* × 10⁻²
cable	meter	2.194 56* × 10²
caliber	meter	2.54* × 10⁻⁴
calorie (International Steam Table)	joule	4.1868
calorie (mean)	joule	4.190 02
calorie (thermochemical)	joule	4.184*
calorie (15° C)	joule	4.185 80
calorie (20° C)	joule	4.181 90
calorie (kilogram, International Steam Table)	joule	4.1868 × 10³
calorie (kilogram, mean)	joule	4.190 02 × 10³
calorie (kilogram, thermochemical)	joule	4.184* × 10³
carat (metric)	kilogram	2.00* × 10⁻⁴
Celsius (temperature)	kelvin	$t_K = t_C + 273.15$
centimeter of mercury (0° C)	newton/meter²	1.333 22 × 10³
centimeter of water (4° C)	newton/meter²	9.806 38 × 10
chain (engineer or ramden)	meter	3.048* × 10
chain (surveyor or gunter)	meter	2.011 68* × 10
circular mil	meter²	5.067 074 8 × 10⁻¹⁰
cord	meter³	3.624 556 3
cubit	meter	4.572* × 10⁻¹
cup	meter³	2.365 882 365* × 10⁻⁴
curie	disintegration/second	3.70* × 10¹⁰
day (mean solar)	second (mean solar)	8.64* × 10⁴
day (sidereal)	second (mean solar)	8.616 409 0 × 10⁴
degree (angle)	radian	1.745 329 251 994 3 × 10⁻²
denier (international)	kilogram/meter	1.00* × 10⁻⁷
dram (avoirdupois)	kilogram	1.771 845 195 312 5* × 10⁻³
dram (troy or apothecary)	kilogram	3.887 934 6* × 10⁻³
dram (U.S. fluid)	meter³	3.696 691 195 312 5* × 10⁻⁶
dyne	newton	1.00* × 10⁻⁵
electron volt	joule	1.602 191 7 × 10⁻¹⁹
erg	joule	1.00* × 10⁻⁷

Table B-2 *(continued)*

To convert from	*to*	*multiply by*
Fahrenheit (temperature)	kelvin	$t_K = (5/9)(t_F + 459.67)$
Fahrenheit (temperature)	Celsius	$t_C = (5/9)(t_F - 32)$
faraday (based on carbon 12)	coulomb	$9.648\ 70 \times 10^4$
faraday (chemical)	coulomb	$9.649\ 57 \times 10^4$
faraday (physical)	coulomb	$9.652\ 19 \times 10^4$
fathom	meter	$1.828\ 8*$
fermi (femtometer)	meter	$1.00* \times 10^{-15}$
fluid ounce (U.S.)	meter³	$2.957\ 352\ 956\ 25* \times 10^{-5}$
foot	meter	$3.048* \times 10^{-1}$
foot (U.S. survey)	meter	$1200/3937*$
foot (U.S. survey)	meter	$3.048\ 006\ 096 \times 10^{-1}$
foot of water (39.2° F)	newton/meter²	$2.988\ 98 \times 10^3$
footcandle	lumen/meter²	$1.076\ 391\ 0 \times 10^1$
footlambert	candela/meter²	$3.426\ 259$
free fall, standard	meter/second²	$9.806\ 65*$
furlong	meter	$2.011\ 68* \times 10^2$
gal (galileo)	meter/second²	$1.00* \times 10^{-2}$
gallon (U.K. liquid)	meter³	$4.546\ 087 \times 10^{-3}$
gallon (U.S. dry)	meter³	$4.404\ 883\ 770\ 86* \times 10^{-3}$
gallon (U.S. liquid)	meter³	$3.785\ 411\ 784* \times 10^{-3}$
gamma	tesla	$1.00* \times 10^{-9}$
gauss	tesla	$1.00* \times 10^{-4}$
gilbert	ampere turn	$7.957\ 747\ 2 \times 10^{-1}$
gill (U.K.)	meter³	$1.420\ 652 \times 10^{-4}$
gill (U.S.)	meter³	$1.182\ 941\ 2 \times 10^{-4}$
grad	degree (angular)	$9.00* \times 10^{-1}$
grad	radian	$1.570\ 796\ 3 \times 10^{-2}$
grain	kilogram	$6.479\ 891* \times 10^{-5}$
gram	kilogram	$1.00* \times 10^{-3}$
hand	meter	$1.016* \times 10^{-1}$
hectare	meter²	$1.00* \times 10^4$
hogshead (U.S.)	meter³	$2.384\ 809\ 423\ 92* \times 10^{-1}$
horsepower (550 foot lbf/second)	watt	$7.456\ 998\ 7 \times 10^2$
horsepower (boiler)	watt	$9.809\ 50 \times 10^3$
horsepower (electric)	watt	$7.46* \times 10^2$
horsepower (metric)	watt	$7.354\ 99 \times 10^2$
horsepower (U.K.)	watt	7.457×10^2
horsepower (water)	watt	$7.460\ 43 \times 10^2$
hour (mean solar)	second (mean solar)	$3.60* \times 10^3$
hour (sidereal)	second (mean solar)	$3.590\ 170\ 4 \times 10^3$
hundredweight (long)	kilogram	$5.080\ 234\ 544* \times 10$
hundredweight (short)	kilogram	$4.535\ 923\ 7* \times 10$
inch	meter	$2.54* \times 10^{-2}$
inch of mercury (32° F)	newton/meter²	$3.386\ 389 \times 10^3$

Table B-2 (*continued*)

To convert from	to	multiply by
inch of mercury (60° F)	newton/meter²	$3.376\ 85 \times 10^3$
inch of water (39.2° F)	newton/meter²	$2.490\ 82 \times 10^2$
inch of water (60° F)	newton/meter²	2.4884×10^2
kayser	1/meter	$1.00^* \times 10^2$
kilocalorie (International Steam Table)	joule	$4.186\ 8 \times 10^3$
kilocalorie (mean)	joule	$4.190\ 02 \times 10^3$
kilocalorie (thermochemical)	joule	$4.184^* \times 10^3$
kilogram mass	kilogram	1.00^*
kilogram force (kgf)	newton	$9.806\ 65^*$
kilopound force	newton	$9.806\ 65^*$
kip	newton	$4.448\ 221\ 615\ 260\ 5^* \times 10^3$
knot (international)	meter/second	$5.144\ 444\ 444 \times 10^{-1}$
lambert	candela/meter²	$1/\pi^* \times 10^4$
lambert	candela/meter²	$3.183\ 098\ 8 \times 10^3$
langley	joule/meter²	$4.184^* \times 10^4$
lbf (pound force, avoirdupois)	newton	$4.448\ 221\ 615\ 260\ 5^*$
lbm (pound mass, avoirdupois)	kilogram	$4.535\ 923\ 7^* \times 10^{-1}$
league (U.K. nautical)	meter	$5.559\ 552^* \times 10^3$
league (international nautical)	meter	$5.556^* \times 10^3$
league (statute)	meter	$4.828\ 032^* \times 10^3$
light year	meter	$9.460\ 55 \times 10^{15}$
link (engineer or ramden)	meter	$3.048^* \times 10^{-1}$
link (surveyor or gunter)	meter	$2.011\ 68^* \times 10^{-1}$
liter	meter³	$1.00^* \times 10^{-3}$
lux	lumen/meter²	1.00^*
maxwell	weber	$1.00^* \times 10^{-8}$
meter	wavelengths Kr 86	$1.650\ 763\ 73^* \times 10^6$
micron	meter	$1.00^* \times 10^{-6}$
mil	meter	$2.54^* \times 10^{-5}$
mile (U.S. statute)	meter	$1.609\ 344^* \times 10^3$
mile (U.K. nautical)	meter	$1.853\ 184^* \times 10^3$
mile (international nautical)	meter	$1.852^* \times 10^3$
mile (U.S. nautical)	meter	$1.852^* \times 10^3$
millibar	newton/meter²	$1.00^* \times 10^2$
millimeter of mercury (0° C)	newton/meter²	$1.333\ 224 \times 10^2$
minute (angle)	radian	$2.908\ 882\ 086\ 66 \times 10^{-4}$
minute (mean solar)	second (mean solar)	$6.00^* \times 10$
minute (sidereal)	second (mean solar)	$5.983\ 617\ 4 \times 10$
month (mean calendar)	second (mean solar)	$2.628^* \times 10^6$
nautical mile (international)	meter	$1.852^* \times 10^3$
nautical mile (U.S.)	meter	$1.852^* \times 10^3$
nautical mile (U.K.)	meter	$1.853\ 184^* \times 10^3$

Table B-2 *(continued)*

To convert from	*to*	*multiply by*
oersted	ampere/meter	$7.957\ 747\ 2 \times 10$
ounce force (avoirdupois)	newton	$2.780\ 138\ 5 \times 10^{-1}$
ounce mass (avoirdupois)	kilogram	$2.834\ 952\ 312\ 5^* \times 10^{-2}$
ounce mass (troy or apothecary)	kilogram	$3.110\ 347\ 68^* \times 10^{-2}$
ounce (U.S. fluid)	meter^3	$2.957\ 352\ 956\ 25^* \times 10^{-5}$
pace	meter	$7.62^* \times 10^{-1}$
parsec (IAU)	meter	$3.085\ 7 \times 10^{16}$
pascal	newton/meter^2	1.00^*
peck (U.S.)	meter^3	$8.809\ 767\ 541\ 72^* \times 10^{-3}$
pennyweight	kilogram	$1.555\ 173\ 84^* \times 10^{-3}$
perch	meter	5.0292^*
phot	lumen/meter^2	1.00×10^4
pica (printers)	meter	$4.217\ 517\ 6^* \times 10^{-3}$
pint (U.S. dry)	meter^3	$5.506\ 104\ 713\ 575^* \times 10^{-4}$
pint (U.S. liquid)	meter^3	$4.731\ 764\ 73^* \times 10^{-4}$
point (printers)	meter	$3.514\ 598^* \times 10^{-4}$
poise	newton second/meter^2	$1.00^* \times 10^{-1}$
pole	meter	5.0292^*
pound force (lbf avoirdupois)	newton	$4.448\ 221\ 615\ 260\ 5^*$
pound mass (lbm avoirdupois)	kilogram	$4.535\ 923\ 7^* \times 10^{-1}$
pound mass (troy or apothecary)	kilogram	$3.732\ 417\ 216^* \times 10^{-1}$
poundal	newton	$1.382\ 549\ 543\ 76^* \times 10^{-1}$
quart (U.S. dry)	meter^3	$1.101\ 220\ 942\ 715^* \times 10^{-3}$
quart (U.S. liquid)	meter^3	$9.463\ 592\ 5 \times 10^{-4}$
rad (radiation dose absorbed)	joule/kilogram	$1.00^* \times 10^{-2}$
Rankine (temperature)	kelvin	$t_K = (5/9) t_R$
rayleigh (rate of photon emission)	1/second meter^2	$1.00^* \times 10^{10}$
rhe	meter^2/newton second	$1.00^* \times 10$
rod	meter	5.0292^*
roentgen	coulomb/kilogram	$2.579\ 76^* \times 10^{-4}$
rutherford	disintegration/second	$1.00^* \times 10^6$
second (angle)	radian	$4.848\ 136\ 811 \times 10^{-6}$
second (ephemeris)	second	$1.000\ 000\ 000$
second (mean solar)	second (ephemeris)	Consult American Ephemeris and Nautical Almanac
second (sidereal)	second (mean solar)	$9.972\ 695\ 7 \times 10^{-1}$
section	meter^2	$2.589\ 988\ 110\ 336^* \times 10^6$
scruple (apothecary)	kilogram	$1.295\ 978\ 2^* \times 10^{-3}$
shake	second	1.00×10^{-8}
skein	meter	$1.097\ 28^* \times 10^2$
slug	kilogram	$1.459\ 390\ 29 \times 10$
span	meter	$2.286^* \times 10^{-1}$
statampere	ampere	$3.335\ 640 \times 10^{-10}$

Table B-2 *(continued)*

To convert from	to	multiply by
statcoulomb	coulomb	$3.335\,640 \times 10^{-10}$
statfarad	farad	$1.112\,650 \times 10^{-12}$
stathenry	henry	$8.987\,554 \times 10^{11}$
statohm	ohm	$8.987\,554 \times 10^{11}$
statute mile (U.S.)	meter	$1.609\,344* \times 10^{3}$
statvolt	volt	$2.997\,925 \times 10^{2}$
stere	meter^3	$1.00*$
stilb	candela/meter^2	1.00×10^{4}
stoke	meter^2/second	$1.00* \times 10^{-4}$
tablespoon	meter^3	$1.478\,676\,478\,125* \times 10^{-5}$
teaspoon	meter^3	$4.928\,921\,593\,75* \times 10^{-6}$
ton (assay)	kilogram	$2.916\,666\,6 \times 10^{-2}$
ton (long)	kilogram	$1.016\,046\,908\,8* \times 10^{3}$
ton (metric)	kilogram	$1.00* \times 10^{3}$
ton (nuclear equivalent of TNT)	joule	4.20×10^{9}
ton (register)	meter^3	$2.831\,684\,659\,2*$
ton (short, 2000 pound)	kilogram	$9.071\,847\,4* \times 10^{2}$
tonne	kilogram	$1.00* \times 10^{3}$
torr (0° C)	newton/meter^2	$1.333\,22 \times 10^{2}$
township	meter^2	$9.323\,957\,2 \times 10^{7}$
unit pole	weber	$1.256\,637 \times 10^{-7}$
yard	meter	$9.144* \times 10^{-1}$
year (calendar)	second (mean solar)	$3.1536* \times 10^{7}$
year (sidereal)	second (mean solar)	$3.155\,815\,0 \times 10^{7}$
year (tropical)	second (mean solar)	$3.155\,692\,6 \times 10^{7}$
year 1900, tropical, Jan., day 0, hour 12	second (ephemeris)	$3.155\,692\,597\,47* \times 10^{7}$
year 1900, tropical, Jan., day 0, hour 12	second	$3.155\,692\,597\,47 \times 10^{7}$

LISTING BY PHYSICAL QUANTITY

ACCELERATION

foot/second^2	meter/second^2	$3.048* \times 10^{-1}$
free fall, standard	meter/second^2	$9.806\,65*$
gal (galileo)	meter/second^2	$1.00* \times 10^{-2}$
inch/second^2	meter/second^2	$2.54* \times 10^{-2}$

AREA

acre	meter^2	$4.046\,856\,422\,4* \times 10^{3}$
are	meter^2	$1.00* \times 10^{2}$

Table B-2 *(continued)*

To convert from	to	multiply by
barn	meter^2	$1.00* \times 10^{-28}$
circular mil	meter^2	$5.067\,074\,8 \times 10^{-10}$
foot^2	meter^2	$9.290\,304* \times 10^{-2}$
hectare	meter^2	$1.00* \times 10^4$
inch^2	meter^2	$6.4516* \times 10^{-4}$
mile^2 (U.S. statute)	meter^2	$2.589\,988\,110\,336* \times 10^6$
section	meter^2	$2.589\,988\,110\,336* \times 10^6$
township	meter^2	$9.323\,957\,2 \times 10^7$
yard^2	meter^2	$8.361\,273\,6* \times 10^{-1}$

DENSITY

gram/centimeter^3	kilogram/meter^3	$1.00* \times 10^3$
lbm/inch^3	kilogram/meter^3	$2.767\,990\,5 \times 10^4$
lbm/foot^3	kilogram/meter^3	$1.601\,846\,3 \times 10$
slug/foot^3	kilogram/meter^3	$5.153\,79 \times 10^2$

ENERGY

British thermal unit:		
(IST before 1956)	joule	$1.055\,04 \times 10^3$
(IST after 1956)	joule	$1.055\,056 \times 10^3$
British thermal unit (mean)	joule	$1.055\,87 \times 10^3$
British thermal unit (thermochemical)	joule	$1.054\,350 \times 10^3$
British thermal unit (39° F)	joule	$1.059\,67 \times 10^3$
British thermal unit (60° F)	joule	$1.054\,68 \times 10^3$
calorie (International Steam Table)	joule	4.1868
calorie (mean)	joule	$4.190\,02$
calorie (thermochemical)	joule	$4.184*$
calorie (15° C)	joule	$4.185\,80$
calorie (20° C)	joule	$4.181\,90$
calorie (kilogram, International Steam Table)	joule	4.1868×10^3
calorie (kilogram, mean)	joule	$4.190\,02 \times 10^3$
calorie (kilogram, thermochemical)	joule	$4.184* \times 10^3$
electron volt	joule	$1.602\,191\,7 \times 10^{-19}$
erg	joule	$1.00* \times 10^{-7}$
foot lbf	joule	$1.355\,817\,9$
foot poundal	joule	$4.214\,011\,0 \times 10^{-2}$
joule (international of 1948)	joule	$1.000\,165$
kilocalorie (International Steam Table)	joule	4.1868×10^3
kilocalorie (mean)	joule	$4.190\,02 \times 10^3$
kilocalorie (thermochemical)	joule	$4.184* \times 10^3$
kilowatt hour	joule	$3.60* \times 10^6$
kilowatt hour (international of 1948)	joule	$3.600\,59 \times 10^6$
ton (nuclear equivalent of TNT)	joule	4.20×10^9
watt hour	joule	$3.60* \times 10^3$

Table B-2   *(continued)*

To convert from	to	multiply by

ENERGY/AREA TIME

Btu (thermochemical)/foot² second	watt/meter²	$1.134\ 893\ 1 \times 10^4$
Btu (thermochemical)/foot² minute	watt/meter²	$1.891\ 488\ 5 \times 10^2$
Btu (thermochemical)/foot² hour	watt/meter²	$3.152\ 480\ 8$
Btu (thermochemical)/inch² second	watt/meter²	$1.634\ 246\ 2 \times 10^6$
calorie (thermochemical)/cm² minute	watt/meter²	$6.973\ 333\ 3 \times 10^2$
erg/centimeter² second	watt/meter²	$1.00^* \times 10^{-3}$
watt/centimeter²	watt/meter²	$1.00^* \times 10^4$

FORCE

dyne	newton	$1.00^* \times 10^{-5}$
kilogram force (kgf)	newton	$9.806\ 65^*$
kilopond force	newton	$9.806\ 65^*$
kip	newton	$4.448\ 221\ 615\ 260\ 5^* \times 10^3$
lbf (pound force, avoirdupois)	newton	$4.448\ 221\ 615\ 260\ 5^*$
ounce force (avoirdupois)	newton	$2.780\ 138\ 5 \times 10^{-1}$
pound force, lbf (avoirdupois)	newton	$4.448\ 221\ 615\ 260\ 5^*$
poundal	newton	$1.382\ 549\ 543\ 76^* \times 10^{-1}$

LENGTH

angstrom	meter	$1.00^* \times 10^{-10}$
astronomical unit (IAU)	meter	$1.496\ 00 \times 10^{11}$
astronomical unit (radio)	meter	$1.495\ 978\ 9 \times 10^{11}$
cable	meter	$2.194\ 56^* \times 10^2$
caliber	meter	$2.54^* \times 10^{-4}$
chain (surveyor or gunter)	meter	$2.011\ 68^* \times 10$
chain (engineer or ramden)	meter	$3.048^* \times 10$
cubit	meter	$4.572^* \times 10^{-1}$
fathom	meter	1.8288^*
fermi (femtometer)	meter	$1.00^* \times 10^{-15}$
foot	meter	$3.048^* \times 10^{-1}$
foot (U.S. survey)	meter	$1200/3937^*$
foot (U.S. survey)	meter	$3.048\ 006\ 096 \times 10^{-1}$
furlong	meter	$2.011\ 68^* \times 10^2$
hand	meter	$1.016^* \times 10^{-1}$
inch	meter	$2.54^* \times 10^{-2}$
league (U.K. nautical)	meter	$5.559\ 552^* \times 10^3$
league (international nautical)	meter	$5.556^* \times 10^3$
league (statute)	meter	$4.828\ 032^* \times 10^3$
light year	meter	$9.460\ 55 \times 10^{15}$
link (engineer or ramden)	meter	$3.048^* \times 10^{-1}$
link (surveyor or gunter)	meter	$2.011\ 68^* \times 10^{-1}$

Table B-2   *(continued)*

To convert from	to	multiply by
meter	wavelengths Kr 86	$1.650\ 763\ 73^* \times 10^6$
micron	meter	$1.00^* \times 10^{-6}$
mil	meter	$2.54^* \times 10^{-5}$
mile (U.S. statute)	meter	$1.609\ 344^* \times 10^3$
mile (U.K. nautical)	meter	$1.853\ 184^* \times 10^3$
mile (international nautical)	meter	$1.852^* \times 10^3$
mile (U.S. nautical)	meter	$1.852^* \times 10^3$
nautical mile (U.K.)	meter	$1.853\ 184^* \times 10^3$
nautical mile (international)	meter	$1.852^* \times 10^3$
nautical mile (U.S.)	meter	$1.852^* \times 10^3$
pace	meter	$7.62^* \times 10^{-1}$
parsec (IAU)	meter	$3.085\ 7 \times 10^{16}$
perch	meter	5.0292^*
pica (printers)	meter	$4.217\ 517\ 6^* \times 10^{-3}$
point (printers)	meter	$3.514\ 598^* \times 10^{-4}$
pole	meter	5.0292^*
rod	meter	5.0292^*
skein	meter	$1.097\ 28^* \times 10^2$
span	meter	$2.286^* \times 10^{-1}$
statute mile (U.S.)	meter	$1.609\ 344^* \times 10^3$
yard	meter	$9.144^* \times 10^{-1}$

MASS

carat (metric)	kilogram	$2.00^* \times 10^{-4}$
gram (avoirdupois)	kilogram	$1.771\ 845\ 195\ 312\ 5^* \times 10$
gram (troy or apothecary)	kilogram	$3.887\ 934\ 6^* \times 10^{-3}$
grain	kilogram	$6.479\ 891^* \times 10^{-5}$
gram	kilogram	$1.00^* \times 10^{-3}$
hundredweight (long)	kilogram	$5.080\ 234\ 544^* \times 10$
hundredweight (short)	kilogram	$4.535\ 923\ 7^* \times 10$
kgf second² meter (mass)	kilogram	$9.806\ 65^*$
kilogram mass	kilogram	1.00^*
lbm (pound mass, avoirdupois)	kilogram	$4.535\ 923\ 7^* \times 10^{-1}$
ounce mass (avoirdupois)	kilogram	$2.834\ 952\ 312\ 5^* \times 10^{-2}$
ounce mass (troy or apothecary)	kilogram	$3.110\ 347\ 68^* \times 10^{-2}$
pennyweight	kilogram	$1.555\ 173\ 84^* \times 10^{-3}$
pound mass, lbm (avoirdupois)	kilogram	$4.535\ 923\ 7^* \times 10^{-1}$
pound mass (troy or apothecary)	kilogram	$3.732\ 417\ 216^* \times 10^{-1}$
scruple (apothecary)	kilogram	$1.295\ 978\ 2^* \times 10^{-3}$
slug	kilogram	$1.459\ 390\ 29 \times 10$
ton (assay)	kilogram	$2.916\ 666\ 6 \times 10^{-2}$
ton (long)	kilogram	$1.016\ 046\ 908\ 8^* \times 10^3$
ton (metric)	kilogram	$1.00^* \times 10^3$
ton (short, 2000 pound)	kilogram	$9.071\ 847\ 4^* \times 10^2$
tonne	kilogram	$1.00^* \times 10^3$

Table B-2   *(continued)*

To convert from	to	multiply by

POWER

To convert from	to	multiply by
Btu (thermochemical)/second	watt	$1.054\ 350\ 264\ 488 \times 10^3$
Btu (thermochemical)/minute	watt	$1.757\ 250\ 4 \times 10$
calorie (thermochemical)/second	watt	4.184^*
calorie (thermochemical)/minute	watt	$6.973\ 333\ 3 \times 10^{-2}$
foot lbf/hour	watt	$3.766\ 161\ 0 \times 10^{-4}$
foot lbf/minute	watt	$2.259\ 696\ 6 \times 10^{-2}$
foot lbf/second	watt	$1.355\ 817\ 9$
horsepower (550 foot lbf/second)	watt	$7.456\ 998\ 7 \times 10^2$
horsepower (boiler)	watt	$9.809\ 50 \times 10^3$
horsepower (electric)	watt	$7.46^* \times 10^2$
horsepower (metric)	watt	$7.354\ 99 \times 10^2$
horsepower (U.K.)	watt	7.457×10^2
horsepower (water)	watt	$7.460\ 43 \times 10^2$
kilocalorie (thermochemical)/minute	watt	$6.973\ 333\ 3 \times 10$
kilocalorie (thermochemical)/second	watt	$4.184^* \times 10^3$
watt (international of 1948)	watt	$1.000\ 165$

PRESSURE

To convert from	to	multiply by
atmosphere	newton/meter²	$1.013\ 25^* \times 10^5$
bar	newton/meter²	$1.00^* \times 10^5$
barye	newton/meter²	$1.00^* \times 10^{-1}$
centimeter of mercury (0° C)	newton/meter²	$1.333\ 22 \times 10^3$
centimeter of water (4° C)	newton/meter²	$9.806\ 38 \times 10$
dyne/centimeter²	newton/meter²	$1.00^* \times 10^{-1}$
foot of water (39.2° F)	newton/meter²	$2.988\ 98 \times 10^3$
inch of mercury (32° F)	newton/meter²	$3.386\ 389 \times 10^3$
inch of mercury (60° F)	newton/meter²	$3.376\ 85 \times 10^3$
inch of water (39.2° F)	newton/meter²	$2.490\ 82 \times 10^2$
inch of water (60° F)	newton/meter²	2.4884×10^2
kgf/centimeter²	newton/meter²	$9.806\ 65^* \times 10^4$
kgf/meter²	newton/meter²	$9.806\ 65^*$
lbf/foot²	newton/meter²	$4.788\ 025\ 8 \times 10$
lbf/inch² (psi)	newton/meter²	$6.894\ 757\ 2 \times 10^3$
millibar	newton/meter²	$1.00^* \times 10^2$
millimeter of mercury (0° C)	newton/meter²	$1.333\ 224 \times 10^2$
pascal	newton/meter²	1.00^*
psi (lbf/inch²)	newton/meter²	$6.894\ 757\ 2 \times 10^3$
torr (0° C)	newton/meter²	$1.333\ 22 \times 10^2$

SPEED

To convert from	to	multiply by
foot/hour	meter/second	$8.466\ 666\ 6 \times 10^{-5}$
foot/minute	meter/second	$5.08^* \times 10^{-3}$
foot/second	meter/second	$3.048^* \times 10^{-1}$
inch/second	meter/second	$2.54^* \times 10^{-2}$

Table B-2   *(continued)*

To convert from	to	multiply by
kilometer/hour	meter/second	$2.777\ 777\ 8 \times 10^{-1}$
knot (international)	meter/second	$5.144\ 444\ 444 \times 10^{-1}$
mile/hour (U.S. statute)	meter/second	$4.4704^* \times 10^{-1}$
mile/minute (U.S. statute)	meter/second	$2.682\ 24^* \times 10$
mile/second (U.S. statute)	meter/second	$1.609\ 344^* \times 10^3$

TEMPERATURE

Celsius	kelvin	$t_K = t_C + 273.15$
Fahrenheit	kelvin	$t_K = (5/9)(t_F + 459.67)$
Fahrenheit	Celsius	$t_C = (5/9)(t_F - 32)$
Rankine	kelvin	$t_K = (5/9)t_R$

TIME

day (mean solar)	second (mean solar)	$8.64^* \times 10^4$
day (sidereal)	second (mean solar)	$8.616\ 409\ 0 \times 10^4$
hour (mean solar)	second (mean solar)	$3.60^* \times 10^3$
hour (sidereal)	second (mean solar)	$3.590\ 170\ 4 \times 10^3$
minute (mean solar)	second (mean solar)	$6.00^* \times 10^1$
minute (sidereal)	second (mean solar)	$5.983\ 617\ 4 \times 10^1$
month (mean calendar)	second (mean solar)	$2.628^* \times 10^6$
second (ephemeris)	second	$1.000\ 000\ 000$
second (mean solar)	second (ephemeris)	Consult American Ephemeris and Nautical Almanac
second (sidereal)	second (mean solar)	$9.972\ 695\ 7 \times 10^{-1}$
year (calendar)	second (mean solar)	$3.1536^* \times 10^7$
year (sidereal)	second (mean solar)	$3.155\ 815\ 0 \times 10^7$
year (tropical)	second (mean solar)	$3.155\ 692\ 6 \times 10^7$
year 1900, tropical, Jan., day 0, hour 12	second (ephemeris)	$3.155\ 692\ 597\ 47^* \times 10^7$
year 1900, tropical, Jan., day 0, hour 12	second	$3.155\ 692\ 597\ 47 \times 10^7$

VISCOSITY

centistoke	meter²/second	$1.00^* \times 10^{-6}$
stoke	meter²/second	$1.00^* \times 10^{-4}$
foot²/second	meter²/second	$9.290\ 304^* \times 10^{-2}$
centipoise	newton second/meter²	$1.00^* \times 10^{-3}$
lbm/foot second	newton second/meter²	$1.488\ 163\ 9$
lbf second/foot²	newton second/meter²	$4.788\ 025\ 8 \times 10$
poise	newton second/meter²	$1.00^* \times 10^{-1}$
poundal second/foot²	newton second/meter²	$1.488\ 163\ 9$
slug/foot second	newton second/meter²	$4.788\ 025\ 8 \times 10$
rhe	meter²/newton second	$1.00^* \times 10$

Table B-2   (*continued*)

To convert from	to	multiply by

VOLUME

acre foot	meter^3	$1.233\ 481\ 837\ 547\ 52* \times 10^3$
barrel (petroleum, 42 gallons)	meter^3	$1.589\ 873 \times 10^{-1}$
board foot	meter^3	$2.359\ 737\ 216* \times 10^{-3}$
bushel (U.S.)	meter^3	$3.523\ 907\ 016\ 688* \times 10^{-2}$
cord	meter^3	$3.624\ 556\ 3$
cup	meter^3	$2.365\ 882\ 365* \times 10^{-4}$
dram (U.S. fluid)	meter^3	$3.696\ 691\ 195\ 312\ 5* \times 10^{-6}$
fluid ounce (U.S.)	meter^3	$2.957\ 352\ 956\ 25* \times 10^{-5}$
foot^3	meter^3	$2.831\ 684\ 659\ 2* \times 10^{-2}$
gallon (U.K. liquid)	meter^3	$4.546\ 087 \times 10^{-3}$
gallon (U.S. dry)	meter^3	$4.404\ 883\ 770\ 86* \times 10^{-3}$
gallon (U.S. liquid)	meter^3	$3.785\ 411\ 784* \times 10^{-3}$
gill (U K.)	meter^3	$1.420\ 652 \times 10^{-4}$
gill (U.S.)	meter^3	$1.182\ 941\ 2 \times 10^{-4}$
hogshead (U.S.)	meter^3	$2.384\ 809\ 423\ 92* \times 10^{-1}$
inch^3	meter^3	$1.638\ 706\ 4* \times 10^{-5}$
liter	meter^3	$1.00* \times 10^{-3}$
ounce (U.S. fluid)	meter^3	$2.957\ 352\ 956\ 25* \times 10^{-5}$
peck (U.S.)	meter^3	$8.809\ 767\ 541\ 72* \times 10^{-3}$
pint (U.S. dry)	meter^3	$5.506\ 104\ 713\ 575* \times 10^{-4}$
pint (U.S. liquid)	meter^3	$4.731\ 764\ 73* \times 10^{-4}$
quart (U.S. dry)	meter^3	$1.101\ 220\ 942\ 715* \times 10^{-3}$
quart (U.S. liquid)	meter^3	$9.463\ 529\ 5 \times 10^{-4}$
stere	meter^3	$1.00*$
tablespoon	meter^3	$1.478\ 676\ 478\ 125* \times 10^{-5}$
teaspoon	meter^3	$4.928\ 921\ 593\ 75* \times 10^{-6}$
ton (register)	meter^3	$2.831\ 684\ 659\ 2*$
yard^3	meter^3	$7.645\ 548\ 579\ 84* \times 10^{-1}$

Piping, heating and ventilating, electrical, and welding symbols

Table C–1 Piping Symbols (ANSI Z32.2.3–1949.)

	FLANGED	SCREWED	BELL & SPIGOT		FLANGED	SCREWED	BELL & SPIGOT
BUSHING				PLUGS			
				BULL PLUG			
CAP				PIPE PLUG			
CROSS				REDUCER			
3.1 REDUCING				CONCENTRIC			
				ECCENTRIC			
3.2 STRAIGHT SIZE				SLEEVE			
CROSSOVER				TEE			
ELBOW							
45-DEGREE				(STRAIGHT SIZE)			
				(OUTLET UP)			
90-DEGREE				(OUTLET DOWN)			
TURNED DOWN				UNION			
TURNED UP							
BASE				ANGLE VALVE			
				CHECK			
DOUBLE BRANCH				GATE (ELEVATION)			
LONG RADIUS				GATE (PLAN)			
REDUCING				GLOBE (ELEVATION)			
SIDE OUTLET (OUTLET DOWN)				GLOBE (PLAN)			
SIDE OUTLET (OUTLET UP)				GLOBE VALVE			
				ANGLE GLOBE	SAME AS	SYMBOLS	15.4 & 15.5
				HOSE GLOBE	SAME AS	SYMBOL	23.3
STREET				MOTOR-OPERATED			
JOINT				HOSE VALVE			
CONNECTING PIPE EXPANSION				ANGLE			
LATERAL				GATE			
				GLOBE			
ORIFICE FLANGE				LOCKSHIELD VALVE			
REDUCING FLANGE				QUICK OPENING VALVE			

Table C–2 Heating and Ventilating Symbols. (ANSI Z32.2.2–1949.)

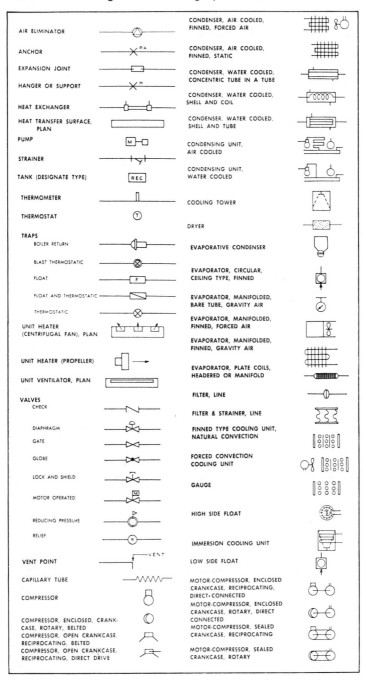

AIR ELIMINATOR	CONDENSER, AIR COOLED, FINNED, FORCED AIR
ANCHOR	CONDENSER, AIR COOLED, FINNED, STATIC
EXPANSION JOINT	CONDENSER, WATER COOLED, CONCENTRIC TUBE IN A TUBE
HANGER OR SUPPORT	CONDENSER, WATER COOLED, SHELL AND COIL
HEAT EXCHANGER	CONDENSER, WATER COOLED, SHELL AND TUBE
HEAT TRANSFER SURFACE, PLAN	
PUMP	CONDENSING UNIT, AIR COOLED
STRAINER	CONDENSING UNIT, WATER COOLED
TANK (DESIGNATE TYPE)	
THERMOMETER	COOLING TOWER
THERMOSTAT	DRYER
TRAPS	
BOILER RETURN	EVAPORATIVE CONDENSER
BLAST THERMOSTATIC	EVAPORATOR, CIRCULAR, CEILING TYPE, FINNED
FLOAT	
FLOAT AND THERMOSTATIC	EVAPORATOR, MANIFOLDED, BARE TUBE, GRAVITY AIR
THERMOSTATIC	EVAPORATOR, MANIFOLDED, FINNED, FORCED AIR
UNIT HEATER (CENTRIFUGAL FAN), PLAN	EVAPORATOR, MANIFOLDED, FINNED, GRAVITY AIR
UNIT HEATER (PROPELLER)	EVAPORATOR, PLATE COILS, HEADERED OR MANIFOLD
UNIT VENTILATOR, PLAN	FILTER, LINE
VALVES	
CHECK	FILTER & STRAINER, LINE
DIAPHRAGM	FINNED TYPE COOLING UNIT, NATURAL CONVECTION
GATE	
GLOBE	FORCED CONVECTION COOLING UNIT
LOCK AND SHIELD	GAUGE
MOTOR OPERATED	
REDUCING PRESSURE	HIGH SIDE FLOAT
RELIEF	IMMERSION COOLING UNIT
VENT POINT	LOW SIDE FLOAT
CAPILLARY TUBE	MOTOR-COMPRESSOR, ENCLOSED CRANKCASE, RECIPROCATING, DIRECT-CONNECTED
COMPRESSOR	MOTOR-COMPRESSOR, ENCLOSED CRANKCASE, ROTARY, DIRECT CONNECTED
COMPRESSOR, ENCLOSED, CRANK-CASE, ROTARY, BELTED	MOTOR-COMPRESSOR, SEALED CRANKCASE, RECIPROCATING
COMPRESSOR, OPEN CRANKCASE. RECIPROCATING, BELTED	
COMPRESSOR, OPEN CRANKCASE. RECIPROCATING, DIRECT DRIVE	MOTOR-COMPRESSOR, SEALED CRANKCASE, ROTARY

Table C–3   Standard Welding Symbols. (American Welding Society)

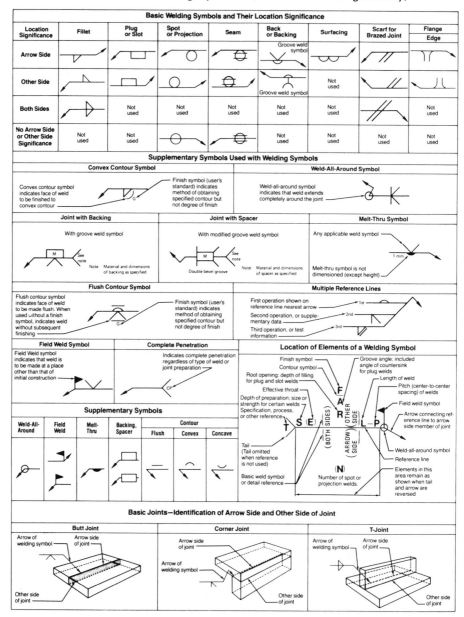

Location Significance	Fillet	Plug or Slot	Spot or Projection	Seam	Back or Backing	Surfacing	Scarf for Brazed Joint	Flange / Edge
Arrow Side					Groove weld symbol			
Other Side					Groove weld symbol	Not used		
Both Sides		Not used	Not used	Not used	Not used	Not used		Not used
No Arrow Side or Other Side Significance	Not used	Not used			Not used	Not used	Not used	Not used

Basic Welding Symbols and Their Location Significance

Supplementary Symbols Used with Welding Symbols

Convex Contour Symbol

Convex contour symbol indicates face of weld to be finished to convex contour

Finish symbol (user's standard) indicates method of obtaining specified contour but not degree of finish

Weld-All-Around Symbol

Weld-all-around symbol indicates that weld extends completely around the joint

Joint with Backing

With groove weld symbol

See note

Note: Material and dimensions of backing as specified

Joint with Spacer

With modified groove weld symbol

See note

Double bevel groove

Note: Material and dimensions of spacer as specified

Melt-Thru Symbol

Any applicable weld symbol

1 mm

Melt-thru symbol is not dimensioned (except height)

Flush Contour Symbol

Flush contour symbol indicates face of weld to be made flush. When used without a finish symbol, indicates weld without subsequent finishing

Finish symbol (user's standard) indicates method of obtaining specified contour but not degree of finish

Multiple Reference Lines

First operation shown on reference line nearest arrow

Second operation, or supplementary data

Third operation, or test information

Field Weld Symbol

Field Weld symbol indicates that weld is to be made at a place other than that of initial construction

Complete Penetration

Indicates complete penetration regardless of type of weld or joint preparation

Location of Elements of a Welding Symbol

Finish symbol

Contour symbol

Root opening; depth of filling for plug and slot welds

Effective throat

Depth of preparation; size or strength for certain welds

Specification, process, or other reference

Tail (Tail omitted when reference is not used)

Basic weld symbol or detail reference

Groove angle; included angle of countersink for plug welds

Length of weld

Pitch (center-to-center spacing) of welds

Field weld symbol

Arrow connecting reference line to arrow side member of joint

Weld-all-around symbol

Reference line

Elements in this area remain as shown when tail and arrow are reversed

(BOTH SIDES)   (ARROW SIDE)   (OTHER SIDE)

Number of spot or projection welds **(N)**

Supplementary Symbols

Weld-All-Around	Field Weld	Melt-Thru	Backing, Spacer	Contour		
				Flush	Convex	Concave

Basic Joints—Identification of Arrow Side and Other Side of Joint

Butt Joint	Corner Joint	T-Joint
Arrow of welding symbol / Arrow side of joint / Other side of joint	Arrow side of joint / Arrow of welding symbol / Other side of joint	Arrow of welding symbol / Arrow side of joint / Other side of joint

Table C–3 (continued)

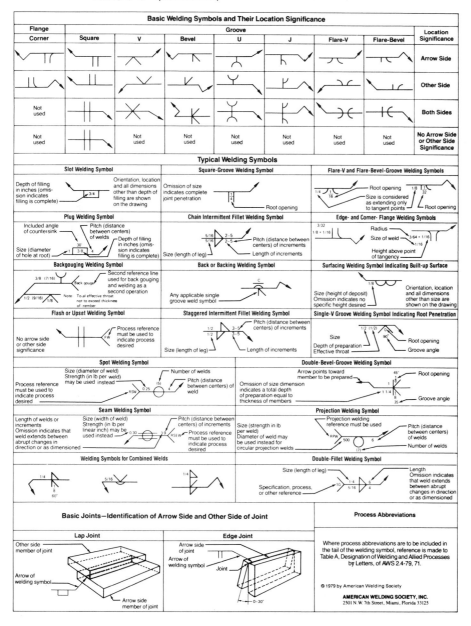

Table C–4 Electrical Symbols. (ANSI) Y32.2–1975.

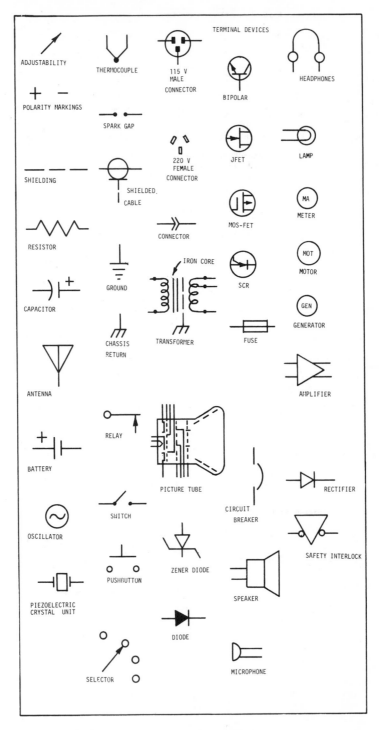

Geometric relationships

D

Table D–1   Areas, Circumferences

Areas

circle	area $= \pi r^2$	$r =$ radius
ellipse	area $= \pi ab$	$a = \frac{1}{2}$ minor axis $b = \frac{1}{2}$ major axis
parallelogram, rectangle	area $=$ bh	$b =$ base, $h \geq$ height
regular polygon	area $= \frac{1}{2}$ snr	$s =$ length of side $n =$ number of sides $r =$ radius of inscribed circle
sphere	area $= \pi d^2$	$d =$ diameter
trapezoid	area $= \frac{1}{2}$ h(b + c)	$b,c =$ two parallel bases
triangle	area $= \frac{1}{2}$ bh	$h =$ height

Circumference (see list above for explanation of terms)

circle	$S = \pi d$
ellipse	S (approx) $= 2\pi \sqrt{(a^2 + b^2)/2}$
rectangle	$S = 2b + 2h$
regular polygon	$S = 2nr \sin(\pi/n)$

Table D–2   Volumes

cone, right circular	$V = (\pi/3)r^2 h$	$r =$ radius, $h =$ height
cylinder, right circular	$V = \pi r^2 h$	
ellipsoid	$V = \frac{4}{3}$ abc	a,b and c are each half of 3 mutually perpendicular axes
prism, right rectangular	$V =$ abc	a,b,c are 3 sides
pyramid	$V = \frac{1}{3}$ bh	$b =$ area of base
sphere	$V = \frac{4}{3} \pi r^3$	
torus	$V = (\pi^2/4)d^2 D$	$d =$ thickness of ring $D =$ diameter of center of ring

D–3 Solid Geometric Figures

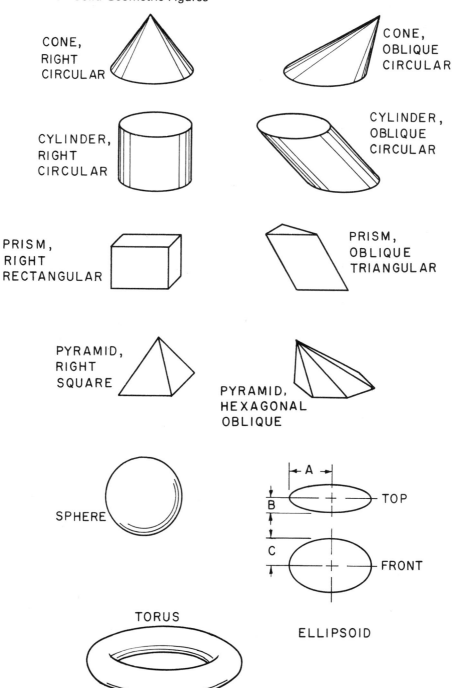

CONE,
RIGHT
CIRCULAR

CONE,
OBLIQUE
CIRCULAR

CYLINDER,
RIGHT
CIRCULAR

CYLINDER,
OBLIQUE
CIRCULAR

PRISM,
RIGHT
RECTANGULAR

PRISM,
OBLIQUE
TRIANGULAR

PYRAMID,
RIGHT
SQUARE

PYRAMID,
HEXAGONAL
OBLIQUE

SPHERE

TOP

FRONT

ELLIPSOID

TORUS

Materials
E

Table E–1 Weight of Metal Sheets

Thickness mm	Weight, kg/m²			
	Steel	Aluminum	Brass	Copper
1	7.85	2.8	8.50	8.90
2	15.70	5.6	17.00	17.80
3	23.55	8.4	25.50	26.70
4	31.40	11.2	34.00	35.60
5	39.25	14.0	42.50	44.50
6	47.10	16.8	51.00	53.40
7	54.95	19.6	59.50	62.30
8	62.80	22.4	68.00	71.20
9	70.65	25.2	76.50	80.10
10	78.50	28.0	85.00	89.00
11	96.35	30.8	93.50	97.90
12	94.20	33.6	102.00	106.80
13	102.05	36.4	110.50	115.70
14	109.90	39.2	119.00	124.60
15	117.75	42.0	127.50	133.50
16	125.60	44.8	136.00	142.40
17	133.45	47.6	144.50	151.30
18	141.30	50.4	153.00	160.20
19	149.15	53.2	161.50	169.10
20	157.00	56.0	170.00	178.00
21	164.85	58.8	178.50	186.90
22	172.70	61.6	187.00	195.80
23	180.55	64.4	195.50	204.70
24	188.40	67.2	204.00	213.60
25	196.25	70.0	212.50	222.50
26	204.10	72.8	221.00	231.40
27	211.95	75.6	229.50	240.30
28	219.80	78.4	238.00	249.20
29	227.65	81.2	240.50	258.10
30	235.50	84.0	255.00	267.00

Table E–2 Weight of Steel Bars
(Density 7.85 g/cm³)

Length mm	Weight, kg/m				Length mm	Weight, kg/m		
	Round	Square	Hex			Round	Square	Hex
1	0.0062	0.0078	0.0068		44	11.9	15.2	13.2
2	0.0247	0.0314	0.0272		45	12.5	15.9	13.8
3	0.0555	0.0706	0.0612		46	13.0	16.6	14.4
4	0.0987	0.126	0.109		48	14.2	18.1	15.7
5	0.154	0.196	0.170		50	15.4	19.6	17.0
6	0.222	0.283	0.245		53	17.3	22.1	19.1
7	0.302	0.385	0.333		55	18.7	23.7	20.6
8	0.395	0.502	0.435		58	20.7	26.4	22.9
9	0.499	0.636	0.551		60	22.2	28.3	24.5
10	0.617	0.785	0.680		63	24.5	31.2	27.0
11	0.746	0.950	0.823		65	26.0	33.2	28.7
12	0.888	1.13	0.979		68	28.5	36.3	31.4
13	1.04	1.33	1.15		70	30.2	38.5	33.3
14	1.21	1.54	1.33		75	34.7	44.2	38.2
15	1.39	1.77	1.53		78	37.5	47.8	41.4
16	1.58	2.01	1.74		80	39.5	50.2	43.5
17	1.78	2.27	1.94		85	44.5	56.7	49.1
18	2.00	2.54	2.20		90	49.9	63.6	55.1
19	2.23	2.83	2.45		95	55.6	70.8	51.4
20	2.47	3.14	2.72		100	61.7	78.5	68.0
21	2.72	3.46	3.00		105	68.0	86.5	
22	2.98	3.80	3.29		110	74.6	95.0	
23	3.26	4.15	3.60		115	81.5	104	
24	3.55	4.52	3.92		120	88.8	113	
25	3.85	4.91	4.25		125	96.3	123	
26	4.17	5.31	4.60		130	104	133	
27	4.50	5.72	4.96		135	112	143	
28	4.83	6.15	5.33		140	121	154	
29	5.19	6.60	5.72		145	130	165	
30	5.55	7.06	6.12		150	139	177	
32	6.31	8.04	6.96		155	148	189	
33	6.71	8.55	7.40		160	158	201	
34	7.13	9.07	7.86		165	168	214	
35	7.55	9.62	8.33		170	178	227	
36	7.99	10.2	8.81		175	189	240	
38	8.90	11.3	9.82		180	200	254	
40	9.87	12.6	10.9		185	211	269	
41	10.4	13.2	11.4		190	223	283	
42	10.9	13.8	12.0		195	234	298	
43	11.4	14.5	12.6		200	247	314	

	Aluminum	Brass	Bronze	Copper	Nickel	Steel	Zinc	Gold
Density g/cm³	2.70	8.36	8.78	8.89	8.8	7.85	7.1	19.3

Table E–3 Characteristics of Seamless Hydraulic Tubing

Nom. Size O.D. × Wall mm	Max. Press. Capacity kPa	Diameter O.D. × I.D. mm	Weight kg/m	Nom. Size O.D. × Wall mm	Max. Press. Capacity kPa	Diameter O.D. × I.D. mm	Weight kg/m
4×0.5	30 700	4×3	0.04	20×2	27 700	20×16	.9
4×0.75	40 100	4×2.5	0.06	20×2.5	34 600	20×15	1.1
4×1	51 200	4×2	0.09	20×3	36 600	20×14	1.3
5×0.75	36 900	5×3.5	0.08	20×3.5	41 800	20×13	1.4
5×1	42 400	5×3	.11	20×4	46 900	20×12	1.5
6×0.75	32 700	6×4.5	.10	22×1	12 600	22×20	.5
6×1.5	53 800	6×3	.18	22×1.5	18 800	22×19	.77
6×2	67 900	6×2	.2	22×2	25 100	22×18	1.0
6×2.5	74 200	6×1.5	.2	22×2.5	27 700	22×17	1.2
8×1	32 700	8×6	.18	22×3	33 100	22×16	1.4
8×1.5	42 300	8×5	.22	25×2	22 200	25×21	1.1
8×2	53 800	8×4	.3	25×2.5	27 700	25×20	1.4
8×2.5	64 500	8×3	.3	25×3	33 100	25×19	1.64
10×1	27 700	10×8	.22	25×4	38 600	25×17	2.05
10×1.5	36 600	10×7	.3	25×4.5	42 900	25×16	2.3
10×2	46 900	10×6	.45	25×5	46 900	25×15	2.45
10×2.5	56 500	10×5	.45	28×1.5	14 800	28×25	.95
12×1	23 000	12×10	.27	28×2	19 700	28×24	1.3
12×1.5	34 600	12×9	.4	28×2.5	24 700	28×23	1.55
12×2	40 100	12×8	.45	28×3	29 600	28×22	1.8
12×2.5	48 500	12×7	.6	28×4	39 500	28×20	2.36
12×3	56 500	12×6	.64	28×5	42 600	28×18	2.8
14×1	19 700	14×12	.3	30×2	18 400	30×26	1.36
14×1.5	29 600	14×11	.45	30×2.5	23 000	30×25	1.7
14×2	39 500	14×10	.6	30×3	27 700	30×24	2
14×2.5	42 600	14×9	.7	30×4	36 900	30×22	2.55
14×3	49 700	14×8	.8	30×5	40 100	30×20	3
14×3.5	56 500	14×7	.86	35×2	15 800	35×31	1.6
15×1	18 400	15×13	.3	35×2.5	19 700	35×30	2
15×1.5	27 700	15×12	.5	35×3	23 700	35×29	2.4
15×2	36 900	15×11	.64	35×4	31 600	35×27	3
15×2.5	40 100	15×10	.77	35×5	39 500	35×25	3.7
15×3	46 900	15×9	.86	35×6	41 000	35×23	4.3
16×1	17 300	16×14	.36	38×2.5	18 200	38×33	2.2
16×1.5	25 900	16×13	.5	38×3	21 900	38×32	2.6
16×2	34 600	16×12	.7	38×4	29 100	38×30	3.3
16×2.5	37 800	16×11	.8	38×5	36 400	38×28	4.
16×3	44 300	16×10	.95	38×6	38 200	38×26	4.7
18×1	15 400	18×16	.4	38×7	43 700	38×24	5.3
18×1.5	23 000	18×15	.64	42×2	13 100	42×38	2
18×2	30 700	18×14	.77	42×3	19 700	42×36	2.9
18×2.5	38 400	18×13	.95	42×4	46 400	42×34	3.7
18×3	40 100	18×12	1.1	50×2		50×46	2.4
20×1.5	20 800	20×17	.7	50×2.5		50×45	2.9
				50×3		50×44	3.45

Outside diameter of the tube indicates the nominal size of the fitting

Maximum safe pressure capacity of fitting

LL—Light pressure   L—Medium pressure   S—Heavy pressure
3900 kPa   9800 kPa   39 000 kPa

Burst point is 50% over maximum pressure

Table E–4 Weight of Flat Steel Bars (Density 7.85 g/cm³)

	B in mm												
A mm	10	12	14	15	16	18	20	25	30	35	40	45	50
1	0.079	0.094	0.110	0.118	0.126	0.141	0.157	0.196	0.235	0.275	0.314	0.353	0.392
2	0.157	0.186	0.220	0.236	0.251	0.283	0.314	0.393	0.471	0.550	0.628	0.707	0.785
3	0.236	0.283	0.330	0.353	0.377	0.424	0.471	0.589	0.705	0.824	0.942	1.060	1.177
4	0.314	0.377	0.440	0.471	0.502	0.565	0.628	0.785	0.942	1.099	1.256	1.413	1.570
5	0.393	0.471	0.550	0.589	0.628	0.707	0.785	0.981	1.177	1.374	1.570	1.766	1.962
6	0.471	0.565	0.659	0.707	0.754	0.848	0.942	1.178	1.413	1.649	1.884	2.120	2.355
7	0.550	0.659	0.769	0.824	0.879	0.989	1.099	1.374	1.648	1.923	2.198	2.473	2.747
8	0.628	0.754	0.879	0.942	1.005	1.130	1.256	1.570	1.884	2.198	2.512	2.826	3.140
9	0.707	0.848	0.989	1.060	1.130	1.272	1.413	1.766	2.119	2.473	2.826	3.179	3.532
10	0.785	0.942	1.099	1.178	1.256	1.413	1.570	1.963	2.355	2.748	3.140	3.533	9.925
11	0.864	1.036	1.209	1.295	1.382	1.554	1.727	2.159	2.590	3.022	3.454	3.886	4.317
12	0.942	1.130	1.319	1.413	1.507	1.696	1.884	2.355	2.826	3.297	3.768	4.239	4.710
13	1.021	1.255	1.429	1.531	1.633	1.837	2.041	2.551	3.061	3.572	4.082	4.592	5.102
14	1.099	1.319	1.539	1.649	1.758	1.978	2.198	2.748	3.297	3.847	4.396	4.946	5.495
15	1.178	1.413	1.649	1.766	1.884	2.120	2.355	2.944	3.532	4.121	4.710	5.299	5.887
20	1.570	1.884	2.198	2.355	2.512	2.826	3.140	3.925	4.710	5.495	6.280	7.065	7.850
25	1.963	2.355	2.748	2.944	3.140	3.533	3.925	4.905	5.888	6.869	7.850	8.831	9.813
30	2.355	2.826	3.297	3.533	3.768	4.239	4.710	5.888	7.065	8.243	9.420	10.60	11.78
40	3.140	3.768	4.396	4.710	5.024	5.652	6.280	7.850	9.420	11.30	12.56	14.13	15.70

Density g/cm³	Aluminum 2.70	Brass 8.36	Bronze 8.78	Copper 8.89	Nickel 8.8	Steel 7.85	Zinc 7.1	Gold 19.3

Anthropometric data

AF

Table F–1   Standing Heights.

Dimension	Description	Percentile		
		10%	*50%*	*90%*
1	Height	1676	1756	1835
2	Eye Height	1546	1625	1702
3	Shoulder Height	1360	1437	1510
4	Elbow Height	1048	1105	1163
5	Wrist Height	801	852	902
6	Knee Cap Height	481	513	548

(All dimensions are in millimeters.)

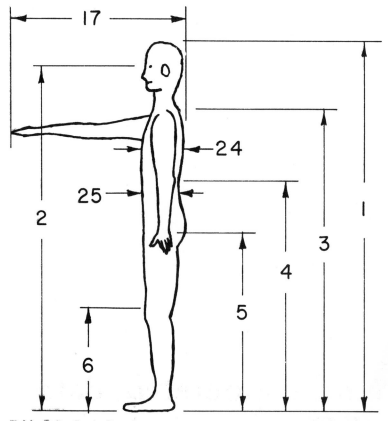

Table F–2   Body Depths.

Dimension	Description	Percentile		
		10%	*50%*	*90%*
24	Chest Depth	207	229	256
25	Waist Depth	176	200	232

Data adapted from a study of 4000 male USAF personnel, not the general population. (Hertzberg, Daniels and Churchill, "Anthropometry of Flying Personnel—1950," WADS Technical Report 52–321, Wright-Patterson Air Force Base, Ohio.)

Table F–3 Sitting Heights.

Dimension	Description	Percentile		
		10%	*50%*	*90%*
7	Sitting Height	872	914	955
8	Shoulder Height	553	592	628
9	Elbow Height	199	232	266
10	Knee Height	126	142	159
11	Thigh Height	519	551	583
12	Popliteal Height	406	431	457
13	Buttock-Knee Length	566	600	634

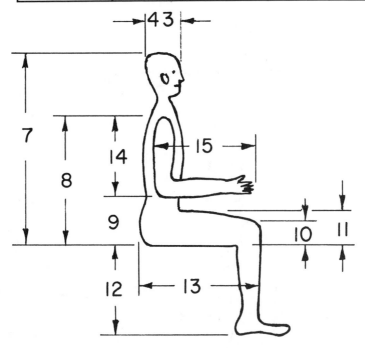

Table F–4 Arm Lengths.

Dimension	Description	Percentile		
		10%	*50%*	*90%*
14	Shoulder-Elbow Length	341	364	387
15	Elbow-To-Finger Length	453	480	506
16	Span	1703	1799	1893
17	Reach	826	879	933

Table F–5   Body Breadths.

Dimension	Description	Percentile		
		10%	50%	90%
18	Elbow-to-Elbow Breadth	395	437	487
19	Hip Breadth	328	354	385
20	Knees Breadth	185	201	219
21	Shoulder Breadth	426	454	485
22	Waist Breadth	243	269	303
23	Hip Breadth	313	334	359

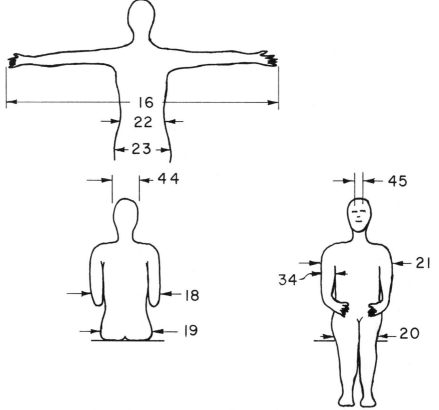

Table F–6   Head Measurements.

Dimension	Description	Percentile		
		10%	50%	90%
43	Length	189	197	206
44	Width	148	154	161
45	Interpupillary	59	63	68
46	Circumference	551	571	592

Table F–7 Hand Measurements.

Dimension	Description	Percentile		
		10%	*50%*	*90%*
39	Length	179	190	201
40	Palm Width	83	88	91
41	Breadth at Thumb	97	101	110
42	Knuckle Thickness	27	30	32

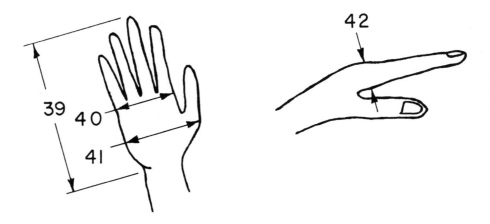

Table F–8 Body Circumferences.

Dimension	Description	Percentile		
		10%	*50%*	*90%*
26	Neck	357	379	404
27	Shoulder	1075	1146	1231
28	Chest	911	982	1069
29	Waist	724	805	919
30	Buttock	888	957	1038
31	Upper Thigh	512	568	626
32	Calf	335	366	398
33	Ankle	210	226	246
34	Biceps	293	324	360
35	Wrist	162	174	187

Table F-9 Foot Dimensions.

Dimension	Description	Percentile		
		10%	50%	90%
36	Length	252	267	281
37	Breadth, Toes	91	96	102
38	Breadth, Heel	62	67	72

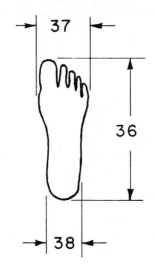

Economic data
G

Table G–1 Cost of selected engineering materials.*

Material	Cost Factor**
Mild Steel	1.0
Stainless Steel	10.7
Copper	10.6
Brass	10.0
Aluminum	8.8
No. 2, yellow pine wood	1.2
No. 3, yellow pine wood	0.6

*These data are presented as guidelines for general comparison. Actual values may vary due to inflation, geographic area, local availability, current market demand, and the particular form supplied.

**To find the actual price of the material, multiply the cost factor by the current cost per unit mass of mild steel.

Table G–2 Relative manpower costs.*

Occupation	Range of Hourly Wage Factor**	Average Hourly Wage Factor**
Engineering Professional	1.60–***	4.48
Technician	1.10–3.07	2.09
Draftsman	1.45–2.64	2.04
Architect	—	3.97
Foundry Worker	1.79–1.83	1.81
Machinist	2.13–2.70	2.41
Tool & Die Maker	1.92–3.06	2.49
Assembler	1.03–2.41	1.72
Welder	1.36–4.14	2.74
Typist	1.22–1.38	1.30
Secretary	1.48–2.34	1.91
Custodian	—	1.25
Construction Worker	1.68–2.59	2.13
Heavy Machine Operator	2.59–3.45	3.02
Plumber-Pipefitter	1.72–3.59	2.66
Ironworker	—	3.45
Electrician	1.72–5.40	3.56

*These data are presented as guidelines for general comparison. Actual values may vary due to inflation, occupational experience, type of industry, geographic area, individual worker quality, and current market demand. These data are adapted from statistics presented in: *Occupational Outlook* 1978–79 Edition, U.S. Dept. of Labor, Bureau of Labor Statistics, 1978, Bulletin 1955.

**To find actual wages in dollars, multiply the listed factor by the present minimum wage.

***Upper bound may vary considerably.

Environmental
design limits
H

Table H–1 Thermal Comfort Chart. (Reprinted with permission from the 1977 Fundamentals Volume, ASHRAE Handbook and Product Directory.)

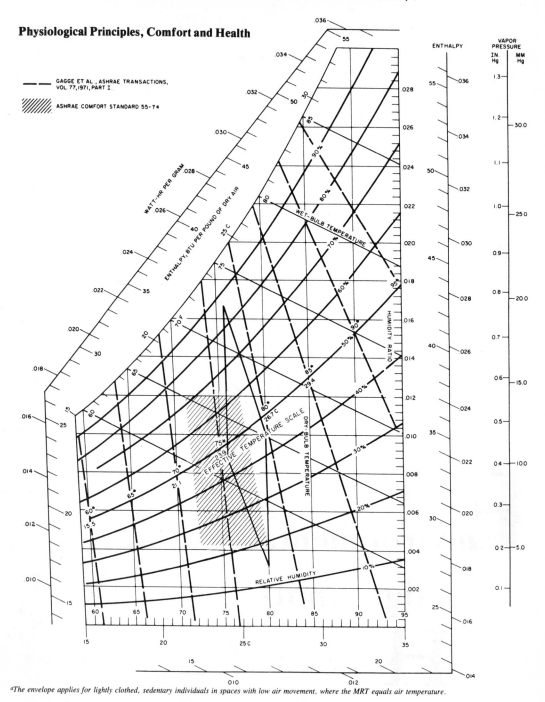

The envelope applies for lightly clothed, sedentary individuals in spaces with low air movement, where the MRT equals air temperature.

Table H–2 Noise Limits*

Noise Level dB A (Slow)	Maximum Permissible Exposure (hr)
90	8
92	6
95	4
97	3
100	2
102	1.5
105	1
110	0.5
115	0.25

*1970 Occupational Safety and Health Administration (OSHA).

NOTES:
[1]First time varying levels, the sum $(C_1/T_1) + (C_2/T_2) + (C_3/T_3) + \ldots$ must not exceed unity. C_1, $C_2 \ldots$ are the actual exposure times to a given level over an 8-hr interval, and T_1, $T_2 \ldots$ are the permitted exposure times given in the Table.

[2]Exposure to impact or impulsive noise should not exceed a peak sound pressure level of 140 dB.

Answers

to selected problems

I

APPENDIX I ANSWERS TO SELECTED PROBLEMS

6.2 J mol^{-1} K^{-1}

6.4 s^{-1}

6.6 dimensionless

6.8 v

6.10 kg m^{-1} s^{-1}

6.12 98.2 N, 90 N, 22.1 lb$_f$, 20.2 lb$_f$

6.14 2.46 m, 279 km hr^{-1}

6.16 1870 J s^{-1}

6.18 7.0 m s^{-1}

6.20 1240 N m^{-2}

6.22 37800

6.24 5.59×10^{-14} N

Index